T0266763

Why Garden in Schools?

This book delves into the complex history of the gardening movement in schools and examines the question why gardens should be built in schools. It offers practical guidance for teachers to begin thinking about how to approach educational gardening.

A resurgence of interest in school gardens is linked to concerns about children's health, food knowledge, lack of outdoor play and contact with the natural world. This book warns against simplistic one-best approaches and makes a case about the complexity of gardening in schools. It is the first critical attempt to address the complex and conflicting notions about school gardens and to tackle the question, 'what is the problem to which school gardens are the answer?' Examining the educational theory in which gardening has been explained and advocated, the book explores the way contemporary garden research has been conducted with specific questions such as 'what works well in school gardens?' Based on case studies of a school establishing a garden and another one maintaining a garden, chapters look at the way in which schools come to frame their gardens. The authors suggest that there are four issues to consider when setting up a school garden or evaluating a pre-existing one – wider social context, public policy, the whole school, and the formal and informal curriculum.

The book ends with a call for consideration of the ways in which school gardens can be built, the myriad practices that constitute an educational garden space and the challenges of maintaining a school garden over the long term. It will be of interest to teachers in primary schools, as well as a key point of reference for scholars, academics and students researching school gardens.

Lexi Earl is a writer and science communicator. She is currently the Outreach and Engagement Manager for the Future Food Beacon at the University of Nottingham. She is the author of *Schools and Food Education in the 21st Century*, published by Routledge in 2018.

Pat Thomson is Professor of Education in the School of Education, the University of Nottingham. Her research agenda is to further understandings about and practices of socially just pedagogies in schools and communities; she often focuses inquiry on the arts and alternative education. She writes, blogs and tweets about academic writing and doctoral education on patthomson.net.

Routledge Spaces of Childhood and Youth Series

Edited by Peter Kraftl and John Horton

The *Routledge Spaces of Childhood and Youth Series* provides a forum for original, interdisciplinary and cutting edge research to explore the lives of children and young people across the social sciences and humanities. Reflecting contemporary interest in spatial processes and metaphors across several disciplines, titles within the series explore a range of ways in which concepts such as space, place, spatiality, geographical scale, movement/mobilities, networks and flows may be deployed in childhood and youth scholarship. This series provides a forum for new theoretical, empirical and methodological perspectives and ground-breaking research that reflects the wealth of research currently being undertaken. Proposals that are cross-disciplinary, comparative and/or use mixed or creative methods are particularly welcomed, as are proposals that offer critical perspectives on the role of spatial theory in understanding children and young people's lives. The series is aimed at upper-level undergraduates, research students and academics, appealing to geographers as well as the broader social sciences, arts and humanities.

Why Garden in Schools?
Lexi Earl and Pat Thomson

Latin American Transnational Children and Youth
Experiences of Nature and Place, Culture and Care Across the Americas
Edited by Victoria Derr and Yolanda Corona-Caraveo

For more information about this series, please visit: https://www.routledge.com/Routledge-Spaces-of-Childhood-and-Youth-Series/book-series/RSCYS

Why Garden in Schools?

Lexi Earl and Pat Thomson

LONDON AND NEW YORK

First published 2021
by Routledge
2 Park Square, Milton Park, Abingdon, Oxon OX14 4RN

and by Routledge
52 Vanderbilt Avenue, New York, NY 10017

Routledge is an imprint of the Taylor & Francis Group, an informa business

British Library Cataloguing-in-Publication Data
A catalogue record for this book is available from the British Library

Library of Congress Cataloging-in-Publication Data
A catalog record has been requested for this book

ISBN: 978-0-367-20758-8 (hbk)
ISBN: 978-0-429-26337-8 (ebk)

Typeset in Times New Roman
by codeMantra

Contents

List of figures vii
List of table ix
Acronyms xi
Acknowledgements xiii

Introduction 1

1 Mapping the history of school gardens 13

2 Contemporary understandings of school gardens 33

3 City School establishes a garden 57

4 New School maintains an established garden 76

5 Analysing the school garden 91

6 Why garden in schools? 110

Index 133

Figures

1.1	Montessori's five purposes for school gardens	18
1.2	Fairview association report	21
1.3	Parsons' pedagogy of the spade	22
1.4	Introduction to the *Nature-study review*, 1905. Issue 1	27
2.1	SAKG evaluation results	42

Table

5.1 Discourses of school gardens 92

Acronyms

ESD	Education for Sustainable Development
SLT	Senior Leadership Team

Acknowledgements

We would like to thank the University of Nottingham, Environment Research Priority Group for providing funding to support our research at City School. We are exceptionally grateful to the two schools we feature in this book for participating in our research. Thanks to the head teachers, school staff and children who spoke to us, and gave up their time to talk with us and share their experiences.

Lexi would like to thank her team at the Future Food Beacon, University of Nottingham, for their support while writing this book. She would also like to thank her family, and particularly Andrés, for their encouragement during this project.

Pat would like to acknowledge the support of her long-suffering partner, Randy, whose prowess in the kitchen grows alongside her book writing.

Introduction

Why garden in schools? Why gardens? Why gardening?

We can easily say why gardens are important to us. We are both keen on gardens. Lexi has an allotment, and Pat has raised beds in her backyard. We plant with hedgehogs, birds and bees in mind. We like to cook with freshly picked seasonal produce. But neither of us are particularly skilled gardeners. Our vegetables would not win prizes in agricultural shows. Our raspberry canes need more expert pruning. We struggle with keeping slugs at bay. We are not qualified horticulturists. So what business do we have in giving any answers to the question of why garden in schools? And why do we think that we can write a book about school gardens?

This introduction outlines our understandings and the approach taken in this book.

School gardening – an idea whose time has come?

We are not alone in taking an interest in gardens.

School gardening is on the increase across the world. Celebrity chefs sponsor high-profile school garden programmes that take a field-to-table approach, promoting the health benefits of buying fresh and local. Government agriculture agencies encourage and support schools to develop gardens that support biodiversity, Indigenous foods and agricultural and horticultural curriculum. Horticultural societies design programmes that advocate gardening as a worthwhile career, leisure activity and learning opportunity. Supermarkets run promotional activities that provide gardening tools and equipment for schools. Development agencies fund school gardens as a poverty alleviation strategy. Conservation groups promote gardening as a path to environmental knowledge and appreciation of nature.

While all of these programmes support a common educational practice – gardening – their different answers to the question 'Why garden in school?' lead to different emphases, different connections with the school curriculum, different kinds of expertise and experts, and different costs to the school. The various rationales also lead to different kinds of research projects with varying claims about the benefits and successes of school gardens. This book canvasses many of these variations but focuses particularly on the school garden as the answer to two different problems: (1) a public health crisis – school gardens prioritise nutrition and cooking but also offer additional cross-curriculum benefits; and (2) an environmental crisis – school gardens are part of a wider 'green' school movement designed to both teach and exemplify Education for Sustainable Development (ESD).

What is a school garden?

We take a broad and inclusive approach to the notion of a garden. The common-sense version of a garden is a cultivated plot of land where ornamental plants and kitchen plants – fruit and vegetables – are grown. However, a garden also generally includes unwanted plants that just grow by themselves – we know these as weeds. There are other garden life forms too – those necessary for germination – bees – and those who use plants as food – butterflies, slugs, snails, beetles and grubs, only some of which are welcome. Gardening might include bee-keeping as it is not only helpful to plant growing but is also a good in its own right. A garden might extend to raising ducks or hens to eradicate insect pests, although each species comes with its own needs, habits and benefits (eggs). This combination of plants and creatures is what most of us have in our yards or allotments, and probably what most people think of when they hear the term 'school garden'.

Ah, but not so fast. A garden is generally a domestic space, but it can also exist at scale. Think of botanic gardens with their educational and conservation goals; plants are carefully categorised and curated for maximum aesthetic and pedagogical effect. Ornamental gardens can also be conservation gardens where Indigenous plants, some rare and endangered, are planted. Experimental gardens might test for the most favourable growing conditions, combinations of nutrients or hardy hybrids. Medicinal gardens specialise in plants with specific healing properties. Sensory gardens allow diverse bodies, senses and sensibilities to engage pleasurably with plants. A school garden might be any of these types of gardens, or incorporate elements of them into the more common vegetable garden. A school garden may actually be multiple spaces which feature elements of all of these, rather than a singular cultivated space.

Perhaps then a school garden is associated with horticulture. Horticulture is a branch of Agriculture and is usually seen as a separate discipline. Some definitions we have seen suggest that Horticulture is enclosed small-scale gardening, while Agriculture is at scale and concerns both food crops and rearing animals[1]. But these definitions are problematic. Apart from the obvious example of large botanic gardens, it is also the case that growing lettuces in a small garden has at least some principles and practices in common with growing them at scale. There is clearly an overlap between Horticulture and Agriculture which may seem semantic, but has obvious potential for internecine disciplinary feuds and compromises within the institutions of further and higher education devoted to growing and rearing.

We take the view here that it is the educational purposes that are most important in school gardens. The gardens might be small or big, draw on the disciplines of Agriculture or Horticulture or not and have a variety of plants and animals in them. Their most important characteristic is the intention to teach children and young people through the medium of a garden – the garden is the 'tool' of a learning programme developed by a school or complementary institution such as alternative provision. The nature of what is seen as 'educational' in the school garden is the focus of this book.

Who are we?

We are educational researchers.

Lexi is the granddaughter of a commercial wheat farmer and grew up in the South African sun. She spent her summers collecting eggs from her grandmothers' hens, and pulling carrots from the earth. She became fascinated by kitchen gardens through her work as a chef, and soon attempted her own growing projects – running a community garden while a PhD student in England and then renting an allotment.

Pat is the child of Depression-era parents who knew the value of self-sufficiency. Her childhood home, a larger version of the Australian quarter-acre block, had fruit trees, grapevines and a variety of vegetables as well as hens. As a younger woman she grew organic vegetables for sale, and has maintained a backyard garden ever since. It has taken her some time to adjust from dry-land gardening to growing in the English climate, with its huge and destructive English slugs and enormous bumbling bees. And as a former school principal of disadvantaged schools in South Australia, she knows something of the practical problems of school gardens.

We have separately researched school garden programmes in the UK, South Africa and Australia. We have encountered conservation gardens used to teach citizenship (Thomson, 2007; Thomson, McQuade, & Rochford, 2005), literacy (Comber, Thomson, & Wells, 2001) and/or environmental sustainability (Thomson, 2006). We have examined feeding programmes, nutrition-oriented gardens (Earl, 2018), gardens for food security (Earl, 2011) and gardens used for healing and mental health (Thomson, Jones, & Hall, 2009; Thomson & Pennacchia, 2016a, b). We have looked at how school gardens are established, managed and maintained (Thomson, Day, Beales, & Curtis, 2010). We have also conducted some research together, which is reported in this book.

Our interest in school gardens extends to a shared theoretical and methodological perspective. This is probably not too surprising as Pat was Lexi's PhD supervisor. This relationship has now morphed into something collaborative, worked out through writing this book together. We are both ethnographers by preference, although we can turn our hand to other methods as needed. We are also both bloggers and take writing seriously. We hope that our book reflects our interest in making difficult ideas accessible.

The approach we take

This book is not simply about gardens and gardening in schools. While we do devote two chapters to two particular school gardens and discuss others, we are really interested in the question of *why*. We are interested in the reasons that schools decide to take up spades, hoes and rakes. We are interested in how they go about the work of designing, planting, weeding and harvesting. We are interested in how the educational possibilities they imagine are materialised, and what students get from their hands-on and hands-in experiences. We are interested in the problems that are encountered. We are interested in whether school gardens are sustained, or are a passing fancy. We are in fact as much interested in exploring the question we pose as its answers.

This book looks at school gardens as solutions to educational problems. We are guided in this approach by Carol Bacchi's work on policies and their interpretation and animation (Bacchi, 1999, 2000, 2009; Bacchi & Bonham, 2014). Bacchi's work is based in Foucault's theorisation of discourse as power-knowledge (Foucault, 1972, 1980). Foucault saw discourse as understanding, thinking, speaking and acting – discourses constitute knowledge, together with allied forms of subjectivity, social practices and power relations. Thinking with

Foucault means examining a field made up of statements and actions, asking, for example, the following questions:

- How were these ways of being, thinking, understanding, relating and acting created and by whom?
- What can be and not be, what can be said and not said, done and not done?
- What is included, left out, foregrounded, backgrounded and obscured?
- How is space made for new and alternative ways of being, thinking, doing, relating and acting?

Bacchi brought Foucauldian theory to the analysis of policy. Bacchi called this the 'What's the problem?' approach, arguing that it is important not to take policy at face value but to ask, 'What is the problem for which this policy is an answer?' Bacchi says that a policy is always a particular formulation of a problem. Two important consequences flow from her argument:

1. Policy problem–posing is based on a selection of information and a set of assumptions. These selections inevitably ignore alternative ways of understanding and posing the problem;
2. The way in which a problem is posed has inbuilt consequences. There are obvious actions – the solution – that need to be taken as a result of the particular problematisation.
   Analysis of any policy text and its enactment thus always involves looking for underpinning problems and their assumptions, omissions, elisions, causalities, inclusions, and blank and blind spots.

An example or two might be helpful here.

A simple problem might be that water from the tap is too hot. It burns. If the problem is understood as resulting from the water rather than the body of the person, the solution is simply to turn on the cold tap and move away from the water until the temperature adjusts. However, imagine that the hot water is at the hairdressing salon. If the problem is understood as being potentially with the person, as well as with the tap, the hairdresser will ask the client whether the water is the right temperature for them, rather than simply using their own preference to adjust the tap.

Of course, most social phenomena are not as simple as scalding water.

Let us take a somewhat more complex issue – too many people are dying in road traffic accidents. Possible problems include careless driving, drivers under the influence of alcohol or drugs, unsafe cars, road visibility and attention-distracting actions such as texting. A range of interventions which attempt to solve the problem can be designed depending on which of the issues seem most pressing – compulsory seat belts, higher safety standards for vehicle building, imposition of alcohol limits and breathalysing, changes to speed limits, changes to driver licensing, public awareness campaigns and so on. One, or a combination, of these problem-solutions inevitably leaves out others – and indeed, we have seen changing government actions designed to address the various problematisations of road safety over time. When one solution seems to only do some of the job, other problems are sought out and remedies are tried. And yet none of these government actions can actually change what individual drivers do, they can only change the driving environment and perhaps public attitudes.

But road safety is a comparatively straightforward matter compared to most social problems and social policies. Educational policy typically deals with very complex matters where one-best, short-term and simplified ways of understanding are not only inadequate but also often have unintended, adverse consequences. The long association of poverty and educational attainment is one such phenomenon. Over time, various problematisations have prevailed. If the poverty-attainment nexus is seen as strongly connected to family income and security, then poverty alleviation, housing and transport policy and employment and industry policy are deemed most important. If lack of basic skills is seen as the most important problem, then the school curriculum is directed towards particular types of literacy teaching, testing and/or mandated time allocated to literacy lessons. If the problem is lack of aspiration, then interventions designed to raise aspirations logically follow.

Bacchi (2000) argues that we 'need to recognise the non-innocence of how problems get framed within policy proposals, how the frames will affect what can be thought about and how this affects the possibilities for action' (p. 30). It is worth adding a food-related example to bring this discussion of discourse and non-innocent problematisation to the heart of the book ahead.

Take the 'problem' of obesity. Obesity has been identified as a problem by the World Health Organisation and governments all around the world. It is framed as a problem of individual failure; that is, people are fat because they choose to be so. Policies reinforce individual responsibility by framing obesity as a problem of energy balance (too many calories in compared to those expended), often even after acknowledging the multiple and complex factors leading to obesity (including genetics, behaviour, environment and culture) (Department of Health, 2017). The policy solutions thus focus on the individual – give people more information on healthy foods as well as promote exercise and healthy choices.

What does such framing do to a problem like obesity? First, it obscures material factors that contribute to the disease – for instance, the proliferation of takeaway shops near school gates and the increasing reliance of children on parent transport rather than walking to school. And, if we think of adult obesity, then technological changes have led to more sedentary office-based work and less work requiring physical labour. Second, a person-centred problematisation obscures the role of industry – the cheapness of processed foods (particularly in comparison to fruits and vegetables), the 2-for-1 deals, the advertising, the addictive nature of ultra-processed foods that encourages people to overeat – all innovations designed to make people want to buy and eat more and more of readily available food products. Framing obesity as a problem of individual failure also leads to fat shaming, body shaming and moral judgments around food choice. Once the policy possibilities for action are framed within the context of individual choice, other actions we might take to fight obesity – for example, holding industry to account around processed foods and their costs – can be glossed over. And people who are obese (the 'out' group in this example) are disempowered by the very solutions meant to help them, and only certain voices in agreement with the individualised problem-solution are heard as meaningful or authoritative on the topic (Ball, 1993).

In this book, following Bacchi, we therefore take the school garden as a solution to a problem that has been explicitly or implicitly posed. We ask, 'What is the problem for which school gardens are seen as the solution?' And we look for the consequences of the specific construction of the problem and solution.

We also add to Bacchi's approach the notions of apparatus and assemblage. An apparatus (*dispositif*) is the term Foucault used to describe how a number of discourses come together in a specific place and time in order to govern behaviours, ways of thinking and being (Foucault, 1972; Peltonen, 2004). We can think of a school as an apparatus (Raffinsoe, Gudmand-Hoyer, & Thaning, 2016) in which discourses around gardens are joined with discourses about pedagogy, child development, citizenship, knowledge and so on. There are always tensions, contradictions and uneasy co-locations as well as strong connections and supports in any discursive apparatus (Foucault, 1980). An apparatus is also mobile, not static, and continually in formation. The closest metaphor we can think of is that a discursive apparatus is a little like tumbleweeds coming together in a larger mass, continually in motion, changing shape and form with the wind. In this metaphor, both the weeds and air are messy, entangled and imbricated, they move or rest together, they are a one. This organic metaphor is much closer to the notion of an assemblage than a machinic apparatus – but this is not the place to engage in this debate. We use the term 'assemblage' in this book, a term which is often taken to be more Deleuzian than Foucauldian, although it is equally a translation of *dispositif*. We understand that there is a complex history to, and argument around, the use of this particular term (e.g. Agamben, 2009; Deleuze, 1992; Legg, 2011; Savage, 2020).

The research we did

There are two portraits of gardens in this book. The first is of a kitchen garden being established at City School. The second is of New School, a garden within an eco-school being maintained and perhaps sustained. These portraits have been developed from more formal case studies.

We were given a small grant by the University of Nottingham to investigate the kitchen garden at City School. Pat and Lexi attended some initial meetings with parents and school staff. Lexi then made two three-day visits to the school. She took photos, and conducted interviews with the head teacher, school cook, administrator and teachers. She also surveyed the children and talked with small groups using photographs and drawings.

The case study of New School was conducted in two separate stages. Together with Bob Curtis, Pat visited New School and interviewed the head and staff members and made a photographic survey of the school. This was one of 12 case studies of school change commissioned by the then National College for School Leadership. Five years later, Lexi did an in-depth ethnographic case study of the school as part of her doctoral research. She conducted extensive observations, worked with maps and drawings and interviewed a range of staff, parents and children. We draw on both sets of data in our portrait of New School as they give us a helpful picture of what happens to a garden over time.

The book that follows

Chapter 1 – Mapping the school garden:

In this chapter we examine what kinds of problems school gardens were historically designed to address. We show how gardens have been taken up in educational theory by Comenius, Rousseau, Froebel,

Pestalozzi and Montessori. We examine the way ideas for children's gardening developed in the late 19th century, and how children were viewed as beings to be 'cultivated'. We explore how the garden emerged as a further space where children could be civilised – moulded and shaped into particular kinds of beings. We examine the early US school gardens of the 20th century. We analyse texts produced about and for 'children's gardening', focussing on the way garden writers wanted to encourage gardening and an interest in the natural world.

Chapter 2 – Contemporary understandings of school gardens:

We consider current scholarship on the histories of school gardens, many of which also use the lens of discourse to consider their purposes and practices. We examine current school garden movements in the US, Australia and England showing the range of benefits and challenges that evaluators identify. Combining the analysis from Chapter 1 with that of Chapter 2 allows us to offer an analytic framing which we use to develop portraits from our case study materials.

Chapter 3 – City School establishes a garden:

This chapter presents the first portrait. We examine a school garden at a primary school in a large city in England. We explore how schools establish a garden, examining the reasoning for undertaking such an endeavour and the various stakeholders involved in the project. We examine how the garden development progresses, the way stakeholders are managed and, crucially, how the school set about funding the garden. We highlight the different ways the garden was framed as a solution to something – health, exposure to nature, food knowledge. We discuss the ways access to the garden was controlled and monitored. We explore the knowledge that children displayed in their conversations with us, and how these contrasted with adult perceptions of child knowledge.

Chapter 4 – New School maintains an established garden:

This chapter offers a second portrait, a school garden with farm elements based at a primary school in a market town in the midlands in England. The school is located in an area of high deprivation. Staff were concerned that children might not be exposed to the natural world within their own home environments. We show how this school garden was established as an integral part of a move to become an 'eco-school' and the way this was driven and promoted by the head teacher. This school featured a farm space, had raised garden beds in different spots around the school, had built a pond, grew cress on the roofs, and had a herb garden. Each of these could be considered an individual 'garden' that together made up a diverse school garden space. We explore the rhetoric and practice of a garden for 'Science' learning, one that allows children to experience the seasons and the natural world, watch lifecycles of mini beasts and 'be' in a garden space. This chapter discusses what happens after an initiating head teacher moves on, taking with them the impetus for outdoor exploration and learning. We talk to the remaining staff and a new head teacher about the future, and discuss the way school priorities shift in the wake of other pressures.

Chapter 5 – Analysing the school garden:

This chapter probes the different discourses we encountered in the school gardens, and the way the problem for which gardening is the solution is tangled in different discursive assemblages. We also focus on the silences of gardening 'talk' – on what is left out of the current conversations around children's gardening. We ask, what is excluded in a school garden? What is left unsaid in garden spaces? How does this define (or not) the schools' identity and their relationships with outsiders and with parents?

Chapter 6 – Why garden in schools?

This chapter considers the challenges of school gardens. First, we examine the curriculum, and the tenuous links that gardens often have with subjects other than health and nutrition and possibly Geography or Science. Second, we consider the range of practical challenges faced by people wishing to establish and maintain school gardens. Finally, we consider questions of school change and key issues of teacher learning, management of resources and school identity and ethos. We conclude the chapter with some unanswered questions about the prospects for school gardening, focussing particularly on whether, and if so how, school gardens might be part of an educational solution to the questions posed by the pandemic and climate collapse. We also add a check list which schools wishing to establish a school garden might consider. Not surprisingly, it begins with the question: 'Why garden in schools?'

A final postscript brings the gardens at New School and City School up to date.

Note

1 http://www.differencebetween.net/miscellaneous/difference-between-agriculture-and-horticulture/

References

Agamben, G. (2009). *What is an apparatus? And other essays.* Stanford, CA: Stanford University Press.

Bacchi, C. (1999). *Women, policy and politics. The construction of policy problems.* London, Thousand Oaks, CA, New Delhi: Sage.

Bacchi, C. (2000). Policy as discourse: What does it mean? Where does it get us? *Discourse, 21*(1), 45–57.

Bacchi, C. (2009). *Analysing policy: What's the problem represented to be?* Frenchs Forest, NSW: Pearson Australia.

Bacchi, C., & Bonham, J. (2014). Reclaiming discursive practices as an anaytic focus. *Foucault Studies, 17*(April), 173–192.

Ball, S. (1993). What is policy? Texts, trajectories and toolboxes. *Discourse, 13*(2), 10–17.

Comber, B., Thomson, P., & Wells, M. (2001). Critical literacy finds a 'place': Writing and social action in a low income Australian grade 2/3 classroom. *Elementary School Journal, 101*(4), 451–464.

Deleuze, G. (1992). What is a dispositif? In T. J. Armstrong (Ed.), *Michel Foucault philosopher* (pp. 159–168). Boston, MA: Harvester Wheatsheaf.

Department of Health. (2017). *Childhood obesity: A plan for action.* London: HM Government.

Earl, A. (2011). *Solving the food security crisis in South Africa: How food gardens can alleviate hunger amongst the poor* (Masters). University of the Witwatersrand.

Earl, A. (2018). *Schools and food education in the 21st century.* London: Routledge.

Foucault, M. (1972). *The archeology of knowledge* (A. Sheridan, Trans. 1995 ed.). London: Routledge.

Foucault, M. (1980). *Power/knowledge: Selected interviews and other writings 1972–1977.* New York: Pantheon Books.

Legg, S. (2011). Assemblage/apparatus: Using Deleuze and Foucault. *Area, 43*(2), 128–133.

Peltonen, M. (2004). From discourse to 'dispositif': Michel Foucault's two histories. *Historical Reflections/Réflexions Historiques, 30*(2), 205–219.

Raffinsoe, S., Gudmand-Hoyer, M., & Thaning, M. S. (2016). Foucault's dispositive: The perspicacity of dispositive analytics in organizational research. *Organization, 23*(2), 272–298.

Savage, G. C. (2020). What is policy assemblage?. *Territory, Policy, Governance, 8*(3), 319–335.

Thomson, P. (2006). Miners, diggers, ferals and showmen: School community projects that unsettle identities? *British Journal of Sociology of Education, 27*(1), 81–96.

Thomson, P. (2007). Making it real: Engaging students in active citizenship projects. In D. Thiessen & A. Cook-Sather (Eds.), *International handbook of student experience in elementary and secondary school* (pp. 775–804). Dordrecht: Springer.

Thomson, P., Day, C., Beales, W., & Curtis, B. (2010 unpublished). *Change leadership. Twelve case studies. NCSL commissioned report.*
Thomson, P., Jones, K., & Hall, C. (2009). *Creative whole school change. Final report.* London: Creativity, Culture and Education; Arts Council England.
Thomson, P., McQuade, V., & Rochford, K. (2005). 'My little special house': Re-forming the risky geographies of middle school girls at Clifftop college. In G. Lloyd (Ed.), *Problem girls. Understanding and supporting troubled and troublesome girls and young women* (pp. 172–189). London: RoutledgeFalmer.
Thomson, P., & Pennacchia, J. (2016a). Discipinary regimes of 'care' and complementary alternative education. *Critical Studies in Education, 57*(1), 84–99.
Thomson, P., & Pennacchia, J. (2016b). Hugs and behaviour points: Alternative education and the regulation of 'excluded' youth. *International Journal of Inclusive Education, 20*(6), 622–640.

1 Mapping the history of school gardens

Maps are imperfect. They are always drawn from someone's point of view. Maps represent the material, cultural and political position of their makers. The history of school gardening we present in this chapter is just like any other map, partial and situated. We are limited by the material available to us. What we present is our interpretation. Our intention is not to simply tell a history of school gardens, as if a straightforward account is both possible and desirable. Rather, we look for the reasons gardens are said to be educationally beneficial. We ask – What is the purpose of a school garden? What is the problem for which school gardening is the answer? What are the consequences of this problematisation? What is omitted? We thus focus on the ways in which the story of school gardens has been told, how gardens have been made important and what is said to be educational about them. We also look for what this story does not say, what connections were not made and what alternative stories might exist, waiting to be told.

The chapter proceeds in two parts. We begin with some educational philosophies which saw the school garden as important. We then move to a 'case' – school gardening history in the US – to raise key questions about learning and school gardens.

School gardens, pedagogies and pedagogues

The school garden backstory often begins in the 17th century with Czech scholar John Amos Comenius, who promoted the idea of a school as a 'garden of delight'. His wholistic imaginary of schools, teaching and curriculum (Comenius, 1907, pp. 111–112) made actual gardens an integral part of the educational experiences on offer. It is perhaps not surprising that in pre-industrial society, where kitchen gardens were a fact of life and teachers were often given land as a means of supplementing their salary (Hemenway, 1915), Comenius understood the potential of a school garden to serve multiple and harmonious purposes. He saw

> no disjuncture between the purposes of providing pleasure, promoting bodily health, inviting reflection and self-examination, turning the mind and heart to God's purposes in the world, promoting economic and cultural development, and seeking social change.
>
> (Smith, 2018, p. 5)

Comenius argued that learning occurs through the senses. Observation, experience and perception led, he said, to retention of knowledge as well as a fresh and permanent memory of the initial event/encounter (Comenius, 1907, p. 185). And what better place to observe and experience than outside in the school garden? Gardens would, Comenius said, give students the opportunity to enjoy and reflect on the natural world, and learn how to care for its plants and animals.

Jean Jacques Rousseau agreed. He also advocated learning through the senses and everyday experiences. The premise of his educational treatise *Emile* (Rousseau, 1889) was that removing a child, and their tutor, from undue influences of family and the wider world, would compel the child to not only learn *how* to survive but also learn *from* how to survive. Children were intrinsically interested in learning, Rousseau suggested, provided the environment was conducive, appropriately organised by the tutor as well as Nature. Rousseau saw a garden as integral to experience-based learning:

> Only twice will it be necessary for him (sic) to see a garden cultivated, seed sown, plants reared, beans sprouting, before he will desire to work in a garden himself.
>
> (Rousseau, 1889, p. 64)

Rousseau's ideas were widely ridiculed at the time of writing. However, his core pedagogical principles have been highly influential over time. Contemporary educators will be familiar with Rousseau's urging of age-appropriate, discovery-led learning; the development of personal accountability through experiencing the consequences of actions taken; the importance of reason over rote memorisation; the importance of learning from life rather than from books; the value of querying authority; and the importance of physical as well as intellectual exercise.

One educator influenced by Rousseau was the Swiss pedagogue Johann Heinrich Pestalozzi. Pestalozzi was an advocate of *anschauung* – impressions gathered through the senses which lead to learning and knowing (Pestalozzi, 1894). Pestalozzi argued that educators build on what children see around them, they use the familiar to teach concepts and categories with their correct names, as well as scientific and moral truths. He believed strongly in Nature as a co-teacher, and favoured domestic and everyday foci for the curriculum, hence a garden (Pestalozzi, 1889). However, unlike Rousseau's wild child, Pestalozzi's students were to begin their tuition in the classroom and then go out into the garden and field to see objects in their natural and changing state. Pestalozzi designed gardens for all of his schools – individual plots as well as a communal plot materialised his philosophy of interconnectedness, unity and individuality.

Pestalozzi's one-time kindred spirit, Frederich Froebel, thought this approach too bookish. 'Doing' was his preferred method. Froebel thought Pestalozzi focussed too much on technical knowledge, stripping away the aesthetic and spiritual from the garden experience. Froebel was particularly interested in younger children and believed that they developed an 'inner need' to read and write at about seven years old. He permitted some earlier mandatory instruction but prohibited writing and reading until the child had reached the appropriate age (Brehony, 1988). Froebel saw the garden as essential to young children's intellectual, physical and moral development.

> For here man (sic) for the first time sees his work bearing fruit in an organic way, determined by the logical necessity and law – fruit which, although subject to the inner laws of natural development, depends in many ways upon his work and upon the character of his work.... if a boy (sic) has given his plants only moderate care and attention, they thrive remarkably well... the plants and flowers of the boys who attend to them with special care live in sympathy with these boys, as it were, and are particularly healthy and luxuriant.
>
> (Froebel, 1908, pp. 111–112)

In other words, the garden reflected the moral attitudes of the child gardener.

Froebel's approach was unpopular at the time. His political opposition to strict Prussian rule was manifest in his non-linear garden design and his experiential pedagogy; his schools were briefly banned for a short period (Herington, 1998). Nevertheless, his naming of early years education as the *kindergarten* (child's garden) has become almost universal, and his influence was widespread. But the divide between Pestalozzi and Froebel – about whether there should be formal instruction prior to hands-on experience and the amount and type of instruction to be given – has continued in education more generally, as well as in relation to garden-based learning programmes.

In Great Britain, Victorian-era educators were heavily influenced by the writings of Herbert Spencer (1855, 1861). Spencer approved of Pestalozzi's arguments for education to follow the processes of the natural world, but saw his ideas as underdeveloped and imperfectly applied. Bringing the 'new' science of psychology and a Darwinian lens to child development, Spencer outlined underpinning educational principles which are still widely held today – learning proceeds from the simple to the complex, thinking becomes more precise as the child matures, lessons should start from the concrete and end in the abstract. Spencer saw that formal schooling had been particularly lax in teaching the knowledge and skills necessary to advance civilisation:

> That increasing acquaintance with the laws of phenomena, which has through successive ages enabled us to subjugate Nature to our need, and in these days gives the common labourer comforts which a few centuries kings could not purchase, is scarcely in any degree owed to the appointed means of instructing our youth. The vital knowledge that by which we have grown as a nation to what we are, and which now underlies our whole existence, is a knowledge that is itself taught in nooks and corners; while the ordained agencies for teaching have been mumbling little else but ideal formulas.
>
> (Spencer, 1861, p. 25)

Spencer's imperial mindset moved away from privileging the Biblical, to prioritising a new spirituality of materialism and science. Education was to position children to be lords over Nature and the world. The school garden became one of many sites where Victorian dame and public schools translated Spencer's philosophy into pedagogy. Applications of Spencer's ideas led to lessons which had children observe and experience nature as competition, as a site of predation by one species on another, as the survival of the best adapted, as a site for the advance of a 'civilising' imperialist nationhood.

Strong advocates for school gardens emerged in tune with the movement for universal school provision. Professor Erasmus Schwab is often credited with establishing a gardening movement in Austria (Gargano, 2007; Jewell, 1907). His writings addressed garden purposes (science, vocation, physical labour) and pedagogies as well as the realpolitik of educational change. Schwab noted disagreements about the purposes of school gardens, insufficient local expertise to design gardens for local conditions, and teachers without either the pedagogical repertoires or subject-based knowledge to design instructional garden programmes. Contemporary school reformers will recognise the challenges he discussed.

> Whoever wishes to make plans for founding suitable school gardens must certainly be an idealist... but he (sic) must also possess the necessary technical knowledge required: he must know life and be acquainted with the public demands by his own inward observations and insight; he must have had intercourse with all classes of the population and must especially be acquainted with teachers, and be himself a school man, in order to be able to meet with question whether his plans can reckon upon general sympathy and furtherance.
>
> (Schwab, 1879, p. 8)

Schwab's rhetoric swept over the difficulties of different student populations and pressures of city and rural schools. Rather than suggesting that gardens might differ by school type, he argued that the primary and universal purpose of the school garden was to teach Natural History and Science. Students should learn about humanity's dependence on nature and its power, and take pleasure in being in and with nature.

As the school garden movement grew, teachers themselves also wrote books about gardens. One example is English school garden practitioner and author Lucy Latter. Latter (1906) was a self-proclaimed Froebelian educator and saw the school garden as a place where experience could be accumulated prior to the formal teaching of reading, writing and arithmetic. Her six-year garden 'experiment', she wrote, not only prepared young children for the work of senior school but also 'withstood the test of Government Inspection' (p. xviii) However, Latter's ideas were strongly critiqued by a highly influential educational thinker, Maria Montessori.

Drawing on a binary rhetoric of practice/practical versus theory/academic well known to contemporary educators, Montessori (1912) argued that Latter's approach was too limited. Latter, she said, focused only on agricultural knowledge and practice in caring for plants, insects and seasons, and domestic notions such as cooking, setting the table and cleaning up. Montessori proposed broad child developmental purposes for school gardens (see Figure 1.1).

Montessori's goals suggest a practice with a strong focus on the social and emotional benefits of gardening, as well as nature appreciation.

Figure 1.1 **Montessori's five purposes for school gardens.**

1 *The child is initiated into observation* of the phenomena of life.
2 The child is initiated into *foresight* by way of *auto-education*; when he (sic) knows that the life of the plants that have been sown depends upon his care in watering them, and that of the animals, upon his diligence in feeding them, without which the little plant dries up and the animals suffer hunger, the child becomes vigilant, as one who is beginning to feel a mission in life.
3 The children are initiated into the virtue of *patience and into confident expectation*, which is a form of faith and of philosophy of life.
4 *The children are inspired with a feeling for nature*, which is maintained by the marvels of creation–that creation which *rewards* with a generosity not measured by the labour of those who help it to evolve the life of its creatures.
5 *The child follows the natural way of development of the human rac*e. (The child is destined to be a civilised man [sic]).

(Montessori, 1912, Chapter 1, italics in original).

Her approach remains influential in contemporary learning gardens as does the secular emphasis on the garden as a civilising experience, inculcating self-discipline, self-management and hard work (see Chapter 2).

The US philosopher John Dewey also wrote about school gardens.

Dewey, a committed advocate of democracy, promoted the virtues of experience for learning how to live and work with others (Dewey, 1897, 1916, 1929, 1934). Dewey was a critic of both book/fact-centred learning and learning led primarily by the child; instead he advocated teaching through the use of structured active inquiries and the use of objects/events/encounters in balance with texts and direct instruction. Gardens directly addressed these possibilities:

> Gardening need not be taught either for the sake of preparing future gardeners, or as an agreeable way of passing time. It affords an avenue of approach to [the] knowledge of the place farming and horticulture have had in the history of the human race and which they occupy in present social organization. Carried on in

> an environment educationally controlled, they [gardens] are a means for making a study of the facts of growth, the chemistry of soil, the role of light, air, moisture, injurious and helpful animal life, etc. There is nothing in the elementary study of botany, which cannot be introduced in a vital way in connection with caring for the growth of seeds. Instead of a subject belonging to a peculiar study called 'botany,' it will then belong to life, and will find, moreover, its natural correlation with the facts of soil, animal life, and human relations.
>
> (Dewey, 1916)

Dewey's ideas were taken up by the US school garden movement which was part of a more general progressive shift in educational philosophy and practice in the US and beyond. Modernisers saw C19th schooling as too tied to a narrow classical curriculum and to memorisation of facts. What was needed, it was held, was an education more fitting with the vision of a modern world. Many progressive educators believed that school gardens could provide a sound basis for a pedagogy where

> the problem grows out of the conditions of the experience being had in the present and it is within the range of the capacity of students: and ... it arouses in the learner an active quest for information and for production of new ideas. The next facts and ideas thus obtained become the ground for further experiences, in which new problems are presented. The process is a continuous cycle.
>
> (Dewey, 1938, p. 79)

But the continuum that Dewey identified, between the child and the subject matter to be taught, and which he saw resolved through the use of experience, inquiry and object, remained a matter for debate. While Dewey advocated gardening as the basis for learning social and scientific knowledge, this was not the only possibility, as was shown in the way the gardening movement developed, as we now explain.

The chapter now turns to study the US school garden movement as a 'case' to show how this mix of philosophical ideas was taken up. We have chosen the US to discuss in part because of the large body of digitally available material, but also because the US represents a systemic approach to school gardening. We return to the discussion of other countries in the next chapter, offering some tentative thinking about why this systemic approach to gardening was not as apparent elsewhere.

The US school gardening movement

Contemporary accounts of the history of US school gardens tell a remarkably similar story – American educators and government officers were influenced by European philosophies of education that included school gardens, and by the efforts of European education systems to bring these philosophies to life. (e.g. Desmond, Grieshop, & Subramaniam, 2004; Hayden-Smith, 2014; Kohlstedt, 2008; Marturano, 1999; United States Department of Agriculture, undated). According to Rose Hayden-Smith (2015) and others (e.g. Burt, 2016; Community of Gardens, undated; Subramaniam, 2002), US school gardens began in earnest in the 1890s in Massachusetts, and then spread as school boards and state and national governments saw their potential. By 1906, 75,000 school gardens were said to exist. This number expanded even further in World War I through the US Garden Army, then declined in number until the School Victory Gardens of World War II. Few were maintained until interest in gardening began to grow again in the 1990s.

One early garden writer, Louise Klein Miller (1908), notes that a body of literature written about gardens, often by women, emerged to support teacher education and teacher practice. It is these literatures we draw on for our case study. These garden literatures illustrate three key themes and tensions that exist around the educational purposes and practices of school gardens – their multiple purposes, as a solution to multiple problems, and possessing multiple connections with the curriculum.

Gardens have multiple purposes

The early 20th-century approach to school gardens was generally broad – this definition from Louise Greene is often cited in today's accounts of school gardening history.

> A school garden may be defined as any garden where children are taught to care for flowers, or vegetables, or both, by one who can, while teaching the life history of the plants and of their friends and enemies, instil in the children a love for outdoor work and such knowledge of natural forces and their laws as shall develop character and efficiency.
>
> (Greene, 1910, p. 1)

School gardens were sites for academic as well as social and moral learning. These mixed mandates were not only derived from European educational philosophers (above) but were strongly in evidence across the US garden movement and its practices (see Figure 1.2).

Figure 1.2 **Fairview association report.**

The school garden's popularity and growth are accounted for in many ways, but chiefly because of its rare combination of essential educational qualities. It is a happy mingling of play and work, vacation and school, athletics and manual training, pleasure and business, beauty and utility, head and hand, freedom and responsibility, corrective and preventive, constructive and creative influences, and all in the great school of out-of-doors. It is a corrective of the evils of the school room. It is a preventive of the perils of misspent leisure. It is constructive of character-building. It is creative of industrious, honest producers. In fact, there is no child's nature to which it does not in some way make a natural and powerful appeal (Livermore, 1908).

Greene saw these multiple purposes in a highly positive light.

> A school garden is like a bank in that it may be drawn upon for values of different kinds to meet different needs. As one may require money in the form of gold or silver, check or draft, in a school garden the educational, economic, aesthetic, utilitarian, or sociological value may be made most prominent, according to circumstance.
>
> (Greene, 1910, pp. 34–35)

While all the US garden literatures we surveyed held strong character-building via gardening as important, there was a marked division between those who focused on the vocational aspects of school gardens, and those who emphasised building disciplinary knowledges. Henry Parsons (1910), for instance, began his book asserting that gardens were to foster the growth of children, not plants (c.f. Montessori v. Latter). Knowledge was, he suggested, the basis for life, health and happiness as well as vital for becoming a member of a well-ordered, wealthy and privileged community and nation (Parsons, 1910, p. 2). Every garden activity thus ought to combine practical activity and academic knowledge – see Figure 1.3.

Alexander Logan (1913) concurred. He argued that the difference between horticulture, where the purpose was only to grow cabbages, and education which stimulated intellectual growth, meant that teachers should be in charge of school gardens, not gardeners as was

Figure 1.3 **Parsons' pedagogy of the spade.**

Digging would teach
... applied physics; economy of effort; development of reason and its application... (as)

1 The three laws of the lever
2 That a small point will enter more easily than a large one
3 The value of keeping the back straight as much as possible in doing the work
4 Economy of personal strength by using the weight when possible not the muscles
5 The value of an instant's relaxation at certain points in a series of movements
6 The individual speed limit.

common practice. By contrast the US Garden Army, speaking at a time when the produce from gardens and student labour was nationally important to food supply, often emphasised the vocational-moral benefits of gardening.

> For the children it will mean health, strength, joy in work, habits of industry, and understanding of the value of money as measured in terms of labor, and such knowledge of the phenomena and forces of nature as must be had for an understanding of most of their school lessons. They will also learn something at least of the fundamental principle of morality, that every man and woman must make his or her own living; must, by some kind of labor of head, hand, or heart, contribute to the common wealth as much as he or she takes from it; must pay in some kind of coin for what he or she gets.
>
> (United States School Garden Army, 1919c, p. 2)

The virtues of hard work and self-reliance, important aspects of the American self-imaginary, clearly paid off – in 1919, over 2 million children were said to have produced some $50 million worth of fruit and vegetables for consumption in their own local communities (United States School Garden Army, 1919b).

Gardens are a solution to multiple problems

Many garden movement authors offered a strong social problem for which school gardens were to provide an answer. A range of ills were presented:

- The city corrupts children and entices them into unproductive if not illegal activities (Clapp, 1903): School gardens were a way to keep children occupied, not only in school time but also after school and on holidays. Gardens would be the means of transporting Nature and the values and benefits of rural life into cities (Hayden-Smith, 2007), with the rural in this context seen idealistically, a romantic and nostalgic idyll. The sudden demise of gardens after the 1920s may be in part explained by the shift of urban populations to the suburbs, where the presence of land for home gardening meant that schools no longer had to compensate for a home deficit (Trelstad, 1997).
- City neighbourhoods are ugly and uncared for: School gardens were a way to provide not only a compensatory aesthetic experience to students but also a community benefit. US school gardener Susan Sipe visited England on a fact-finding mission and reported being

  > impressed with the earnest belief of the Whitechapel teachers in the immense value of nature study in the education of the Jewish and Russian children whom they are training for British subjects, with the spirit of self-sacrifice in these underpaid men and women, and with the intelligent enthusiasm shown by the children into whose lives comes nothing of the beautiful except that brought by the teachers, and whose natural environment is smoke, mud, and rain.
  >
  > (Sipe, 1909, p. 7)

  Residents and passers-by would not only appreciate the beauty of the school grounds but would also see an exemplar of what could be achieved. Children were often sent home from school with cuttings and plants to begin to beautify their out-of-school environment and to involve their families in gardening.
- City children do not get enough exercise and fresh air (Latter, 1906): Gardening required children to be outside and engaged them in productive physical activity, which also had 'moral and mental' benefits. While most schools employed specific gardening staff, they relied heavily on the labour of children, usually boys,

as this anonymous extract from a 1913 publication about school gardening in Los Angeles shows:

> Two thousand boys from the fifth to the eight grade included, devote one and one-half hours each week to gardening under the supervisor and principal, while the girls of the corresponding grades sew.
>
> (Anon, 1913, p. 151)

- Because gardening is a practical skill, parents who did not see the point of school might support them staying on rather than going to work (the child labour problem) (Jarvis, 1916): Early leaving would be prevented because school gardens were relevant and not bookish and removed from children's lives and futures. As well, gardening in school might head off the exodus of young people from rural areas and encourage city children to move to the suburbs where they could have a plot of their own (Miller, 1904, p. 7). Community-based horticultural knowledges, although not those of First Nations peoples, might also be brought into school to enhance the meaningfulness of school gardens, although these knowledges always needed testing out.

> Little Dick's father has told him that back East they plant corn in hills, and Dick is allowed to plant his corn in hills. And seeing how much harder it is to keep his corn well irrigated and how much more slowly it grows than the girl's corn in the next row, he learns that some things that are all right back East won't do in California.
>
> (Waggoner, 1915, p. 215)

Much of the gardening literature was addressed to teachers of Little Dicks. Writers took a highly gender-normative view, in line with society more generally. Boys were assumed to be able to do the heavy work of digging, hoeing and raking; girls were seen as more suited to weeding, pruning and harvesting. There were, of course, some exceptions. The UK Froebelian Lucy Latter, influential in the US, advocated that a girl and boy share a plot for the year (Latter, 1906, p. 9). And H. D. Hemenway in the US saw gardening as 'especially valuable to girls who do not have the same liberties of the street as boys, and are in the open air and sunshine all too little for good, strong physical development' (Hemenway, 1915, p. xvi).

- Gardening was particularly useful for troubling and troublesome groups of children: The terminology used to categorise children who could benefit from gardens, and the characteristics attributed to them, is shocking to today's educators. Miller (1908) describes the successes achieved by dividing the work of gardening into that suitable for 'normal, defective and delinquent' children. She describes the benefits of a garden project with one student Charlie.

> When the subject of a garden was proposed to the students, I was greeted with satisfaction and appreciation, but except one boy. He told me he is not able to work. Charlie is an Italian. His lower limbs are withered and crippled and he walked on two crutches. He has been refused admission to every institution in the state, and sent from the public schools on account of his violent fits of temper. He was made superintendent of the garden. All directions are given to him and he sees that they are executed. He is intelligent and efficient and will secure good results. The whole expression of his face has changed.
>
> (Miller, 1908, p. 579)

Despite the terminology, contemporary educators will not be surprised that gardening was advocated as a solution for children that schools routinely failed. The labour of gardening and the cultivation of plants was (and is still, as we will argue) a form of pedagogic proprietary medicine, a panacea for a wide array of social deficits. The redemptive powers of school gardens are made explicit in much of the historical literature.

Gardens were proposed as solutions to problems with complex histories and bitter political struggles – even the consequences of enslavement and genocide were seen as amenable to a gardening solution. Gardening in school was said to resolve 'the problems of the idle Negro… living on the outskirts of cities and small towns' (United States School Garden Army, 1919c, p. 3). The cruel irony of advocating gardening for the peoples whose labour produced the cotton and tobacco that made the US economically secure was not obvious to school garden enthusiasts. Nor was the culpability of schooling in the ghettoisation, unemployment and poverty they thought school gardens would redress.

Gardens have multiple curriculum connections

Gardens were not only advocated as sites for personal and moral development, they were also a place where children might learn the skills of observation (Logan, 1913), the scientific practices of experimentation and artisan practices of crafting (Brewer, 1913), and various school subjects – Geography, Arithmetic, Drawing, Music, Domestic Science and manual training (Miller, 1908), Economics (Jones, 1923), language and business learning (Hemenway, 1915). Here we address two garden learning domains that remain important today – Horticulture and Science.

Horticulture

Gardening was seen as a body of knowledge in its own right. Teachers and garden supervisors were provided with courses of study for different year groups (United States School Garden Army, 1919b), for children in cities and rural areas (United States School Garden Army, 1919c), and for different parts of the country with different climatic conditions (United States School Garden Army, 1919a). Year-long programmes offered a monthly sequence of garden activities, building knowledge in tune with the seasons (Weed & Emerson, 1909). Detailed advice was provided about what kinds of gardens should be provided in each and every school:

> Each school should be supplied with space for a lawn, a wild garden, a small formal garden, a nursery of trees and shrubs, a vegetable garden, and a small glass house beside the playground. When ample provision for these has been made and the practical work is under the direction of a trained supervisor, it can be properly systematised and graded, stated periods of the school-time being set apart for regular work in the garden.
>
> (Miller, 1904, p. 5)

Horticultural support sometimes took the form of sample designs and planting layouts together with advice on the amount of time children should spend in the garden (Harbor, 1911; c.f. Latter, 1906; Schwab, 1879). Advice was given on how to manage individual plots and communal shared plantings. While there was a general consensus that a garden made up solely of individual plots led to much wasted and weedy space taken up by pathways, it seemed that motivation was often lost in communal gardens. The answer was to combine individual rows with some shared space – to create efficient use of space while encouraging 'healthy rivalry', as well as social and cooperative effort (Brewer, 1913; Waggoner, 1915).

Science

There are frequent references in school garden publications of the early 20th century to the garden as a place for Science learning, nature appreciation, Nature Science and Nature Study, as well as Agriculture and Horticulture. Yet the histories of US school gardens do not adequately address the separate Nature Study movement, which was not necessarily concerned with gardens. The formation of the US Nature Study movement predated school gardens. The Nature Study movement was largely led by scientists (Kohlstedt, 2005) concerned to connect science with everyday life.

Nature Science proponents wanted to move science out of laboratories and into classrooms, to enliven science in ways that would resonate with children and young people. This desire was embedded in wider social changes – according to Armitage (2009), 'Progressive Era Americans' became 'increasingly uneasy about the dispassionate character of social and economic life' in an emerging industrial society. They turned to 'nature for unmediated experiences that might enhance the joy of life'. Armitage argues that Nature Study in schools produced a conservation ethic which influenced some of the great C20th ecologists such as Rachel Carson.

Nature Study educators drew on the same body of European thinking as did school garden proponents – Comenius, Rousseau, Pestalozzi, Froebel, Montessori (Minton, 1980). But Nature Study was specifically focused on the inculcation of scientific knowledge and the aesthetic-spiritual value of nature appreciation (Bailey, 1909; Kass, 2018). Nevertheless, other learnings were also acknowledged (see Figure 1.4).

Figure 1.4 **Introduction to the *Nature-study review*, 1905. Issue 1.**

> Nature study ... includes all the 'natural science' studies of the lower school: the natural history of plants and animals (nature study in its common and most limited sense), school gardening and the closely allied elementary agriculture, elementary physical science, the physical side of geography and physiology and hygiene with special reference to the human body.
>
> (p. 1)

Practices included:

> (1) elementary agriculture; (2) simple object lessons on plants and animals, (3) informal teaching about natural things seen by pupils, for the sake of developing interest and habits of observing; (4) serious elementary biology and physical science; (5) popular picnics in the woods; (6) sentimental talks and reading about plants and animals; (7) 'teaching children to love Nature'.
>
> (p. 3)

However, the scientific community were neither uniformly supportive nor in agreement about the purposes and practices of Nature Study and its connections with fundamental sciences such as Botany, Chemistry and Physics.

School garden advocates often represented Nature Study as somewhat abstract (a different instance of the practical versus abstract binary). Logan (1913), for instance, depicted Nature Study as 'wholly observational and acquisitional', whereas 12–13-year-olds needed additional practical applications and opportunities to develop problem-solving, observation and experimentation. School gardens *could* be integral to Nature Study, advocates argued. Gardens could provide a contained exemplar of the life cycle as plants were germinated, flowered, fruited and died. Gardens provided opportunities to appreciate the beauty and wonder of Nature. They were also a place to exercise control over Nature via cultivation, directing the powers of the natural world to human needs and interests – becoming civilised.

> When nations ceased to be savage, gave up wandering and settled down, they had to make gardens for food.
>
> (French, 1914, p. 1)

But the colonising garden perspective, frequently strong in school garden writings, may not have been attractive to Nature Study proponents who held strong beliefs about wilderness and conservation.

We suspect that the key to the long life of Nature Study in primary schools was that it was directly connected to a school subject – Science. School gardens did not have such an easy link. The two key exceptions seem to be:

1. when Horticulture or Agriculture were offered as vocational subjects and connected with further and higher education – in the US and also in the Macdonald garden movement in Canada (Canadian Heritage Matters, 2018) and
2. when they were needed for the national war effort (Gabler, 1942).

Garden advocates saw the lack of connection between gardens and a single core subject syllabus as a problem. Teachers were often reluctant to add gardening to their existing subject-based work. Enthusiast teachers were thus key to gardening success. As one garden writer noted,

> Some teacher – it may be a science teacher if you can prevail upon him (sic) to deviate from the course of study in general science or biology – should plan with the pupils their gardening adventure.
>
> (Gabler, 1942, p. 470)

We return to the question of teacher expertise and curriculum connection in the last chapter, noting in conclusion to our US school garden history survey that the ideal of the school garden as a site for multiple and various learnings recurs throughout its recorded European history.

Moving forward

We are not simply concerned in this book with the many educational goals loaded onto one garden site, but also with their varying combinations and emphases. We saw in our readings of historical materials that tensions between goals were rarely a recognised part of the thinking – instead, the multiple purposes shown in Comenius' manifesto for education continued. We also saw that intentions, rather than consequences, were a feature of much of the thinking about education through gardens.

But we have taken from these historical literatures, and carry into our further reading and portraits, the following questions:

- What is the particular combination of purposes for the garden? What problems does the garden solve?
- What kind of learning does a school garden support? What pedagogies do teachers use?
- How is the garden organised in time/space? What is the garden – what is important about it (size, plants location, etc.)? What might the garden materialise?
- What school subjects is the garden connected with? What is the disciplinary basis of teaching?
- What successes and problems are encountered in establishing and sustaining the gardens?
- How are human–Nature relationships conceived and enacted?

References

Anon. (1913). School gardens. *The Journal of Education, 77*, 151.

Armitage, K. C. (2009). *The nature study movement. The forgotten popularizer of America's conservation ethic.* Lawrence: University Press of Kansas.

Bailey, L. H. (1909). *The nature-study idea. An interpretation of the new school-movement to put the young into relation and sympathy with nature.* New York: The Macmillan Company.

Brehony, K. J. (1988). *The Froebel movement and state schooling 1880–1914: A study in educational ideology.* (PhD). Open University Milton Keynes.

Brewer, G. W. S. (1913). *Educational school gardening and handwork.* Cambridge: Cambridge University Press.

Burt, K. G. (2016). A complete history of the social, health and political context of the school gardening movement in the United States, 1840–2014. *Journal of Hunger & Environmental Nutrition, 11*(3), 297–316.

Canadian Heritage Matters. (2018). The greatest education – Nature: School gardening in Canada. https://canadianheritagematters.weebly.com/heritage--history/the-greatest-education-nature-school-gardening-in-canada

Clapp, H. (1903). School gardens, city school yards, and the surroundings of rural schools. *The Journal of Education, 58*(4), 80.

Comenius, J. A. (1907). *The great didactic. Setting forth the whole art of teaching all things to all men* (M. W. Keatinge, Trans.). London: Adam and Charles Black.

Community of Gardens. (undated). *Growing from the past: A short history of comunity gardening in the United States: School gardens.* https://communityofgardens.si.edu/exhibits/show/historycommunitygardens/schoolgardens: Smithsonian Museum

Desmond, D., Grieshop, J., & Subramaniam, A. (2004). *Revisiting garden-based learning in basic education.* Paris: International Institute for Educational Planning, UNESCO.

Dewey, J. (1897). My pedagogic creed. *The School Journal, LIV*(3), 77–80.

Dewey, J. (1916). *Democracy and education. An introduction to the philosophy of education* (1996 ed.). New York: Free Press.

Dewey, J. (1929). *Experience and nature* (1958 ed.). New York: Dover.

Dewey, J. (1934). *Art as experience* (1980 ed.). New York: Perigee.

Dewey, J. (1938). *Experience and education* (1963 ed.). New York: Collier Books.

French, A. (1914). *The beginner's garden book. A textbook for the upper grades.* London: The Macmillan Company.

Froebel, F. (1908). *The education of man* (W. N. Hailman, Trans.). New York: D. Appleton and Company.

Gabler, E. R. (1942). School gardens for victory. *The Clearing House, 16*(8), 469–472.

Gargano, E. (2007). *Reading Victorian schoolrooms. Childhood and education in nineteenth century fiction.* Abingdon: Routledge.

Greene, M. L. (1910). *Among school gardens.* New York: Russell Sage Foundation.

Harbor, J. (1911). *Rittenhouse school and gardens.* Toronto: William Briggs.

Hayden-Smith, R. (2007). 'Soldiers of the soil'; the work of the United States GArden Army during World War 1. *Applied Environmental Education and Communication, 6*(1), 19–29.

Hayden-Smith, R. (2014). *Sowing the seeds of victory. American gardening programs of World War 1*. Jefferson, NC: McFarland & Company, Inc.
Hayden-Smith, R. (2015). A history of school gardens ... and how the model is getting a boost today from Foodcorps. *UC Food Observer*, May 6.
Hemenway, H. D. (1915). *How to make school gardens*. New York: Doubelday, Page & Company.
Herington, S. (1998). The garden in Froebel's kindergarten: Beyond the metaphor. *Studies in the History of Gardens & Designed Landscapes, 18*(4), 326–338.
Jarvis, C. D. (1916). *Gardening in elementary schools*. Washington, DC: Bureau of Education.
Jewell, J. R. (1907). *Agricultural education including nature study and school gardens*. Washington, DC: Department of the Interior Bureau of Education.
Jones, R. G. (1923). School gardens. *The Journal of Education, 97*(21), 570.
Kass, D. (2018). *Framing the environmental humanities*. Leiden: Brill.
Kohlstedt, S. (2005). Nature, not books: Scientists and the origins of the nature-study movement of the 1890s. *Isis, 96*(3), 324–352.
Kohlstedt, S. (2008). A better crop of boys and girls: The school gardening movement 1890–1920. *History of Education Quarterly, 48*(1), 58–93.
Latter, L. (1906). *School gardening for little children*. Bloomsbury: Swan Sonnenschein & Co. Lim.
Livermore, A. L. (1908). *School gardens. Report of the Fairview Garden School Association, Yonkers, NY.* Yonkers, NY: Fairview Garden School Association. https://babel.hathitrust.org/cgi/pt?id=uiuo.ark:/13960/t8jd9qs10&view=1up&seq=1
Logan, A. (1913). *Principles and practice of school gardening*. London: Macmillan and Co.Limited.
Marturano, A. (1999). The educational roots of garden based instruction and contemporary gateways to gardening with children. *Kindergarten Education: Theory, Research, Practice, 4*(1), 55–70.
Miller, L. K. (1904). *Children's gardens for school and home. A manual of cooperative gardening*. New York: D. Appleton and Company.
Miller, L. K. (1908). School gardens. *The Elementary School Teacher, 8*(10), 576–580.
Minton, T. G. (1980). *The history of the nature-study movement and its role in the development of environmental education*. University of Massachusetts Amherst, https://scholarworks.umass.edu/dissertations/AAI8019480/
Montessori, M. (1912). *The Montessori method. Scientific pedagogy as applied to child education in 'The Children's Houses' with additions and revisions by the author* (A. E. George, Trans.). New York: Frederick A Stokes Company.
Parsons, H. G. (1910). *Children's gardens for pleasure, health and education*. London: The Macmillan Company.
Pestalozzi, J. H. (1889). *Leonard and Gertrude* (E. Channing, Trans.). Boston, MA: D.C.Heath & Company.
Pestalozzi, J. H. (1894). *How Gertrude teaches her children. An attempt to help mothers teach their children* (H. L. A & F. C. Turner, Trans.). Syracuse, NY: Swan Sonneschein & Co.

Rousseau, J. J. (1889). *Emile, or concerning education* (E. Worthington, Trans.). Boston, MA: D.C. Heath & Company.
Schwab, E. P. (1879). *The school garden. A practical contribution to the subject of education* (M. H. Mann, Trans.). New York: M.L.Holbrook & Co.
Sipe, S. (1909). *School gardening and nature study in English rural schools and in London.* Washington, DC: US Department of Agriculture.
Smith, D. I. (2018). Schools, ideals, gardens. *International Journal of Christianity and Education, 22*(1), 3–7.
Spencer, H. (1855). *The principles of psychology.* London: Longman, Brown, Green, and Longman.
Spencer, H. (1861). *Education. Intellectual, moral, and physical.* London: Willams and Norgate.
Subramaniam, S. (2002). *Garden-based education in basic education: A historical review*, http://fourhcyd.ucdavis.edu: UC Davis Centre for Youth Development.
Trelstad, B. (1997). LIttle machines in their gardens: A history of gardens in America 1891–1920. *Landscape Journal, 16*(2), 161–173.
United States Department of Agriculture. (undated). The school garden.
United States School Garden Army. (1919a). *Forty lessons in gardening for the northeastern states.* Washington, DC: Bureau of Education.
United States School Garden Army. (1919b). *Home gardening for city children of the fifth, sixth and seventh grades.* Washington, DC: Bureau of Education.
United States School Garden Army. (1919c). *Home gardening for town children.* Washington D.C: Bureau of Education.
Waggoner, E. (1915). Los Angeles school gardens. *The Journal of Education, 8*(18), 214–215.
Weed, C. M., & Emerson, P. (1909). *The school garden book.* New York: Charles Scribener's Sons.

2 Contemporary understandings of school gardens

In the past, school gardens were rarely researched. As 21st-century researchers oriented to evidence and impact, we were a little shocked to see the assumptions and assertions made in the historical literatures we read. Most authors that we encountered simply stated the benefits of gardening without recourse to any form of data. Some used anecdotes and unsystematic narrative case studies. Percentages were used without any reference to the source or the method used to generate them. An exception to this rule appeared to be the raw numbers of students involved in programmes. Our shock, of course, says as much about our positioning and socialisation as researchers as it does about the histories, all written by people with years of experience and considerable professional knowledge.

Today's school garden programmes are much more likely to be researched. Funders routinely demand evaluations in order to assess the success or otherwise of their intervention, and to decide whether a programme is worth continuing. Researchers interested in policy and practice are often keen to go further than funded evaluations, and conduct more extensive investigations. There is thus a body of contemporary literature on school gardens that we can draw on, not only to help us think about our portraits but also to orient our overall analysis.

This chapter surveys current research literatures on school gardens. It begins by briefly looking at some contemporary school garden histories. It then moves to studies which examine current school gardening programmes, noting the difference between the interest of the historical researchers in larger social questions of race, gender and nation and much of the contemporary 'what works and how well' evaluations. The chapter concludes by examining an area where there is relatively little research published, that of the school garden programme as a school change initiative. The chapter sets the scene for the two ethnographic whole school portraits that follow.

Writing today – themes from contemporary school garden histories

Today's educational historians have produced overviews of the school garden movement, as well as some detailed studies of individual schools and localities. These accounts allowed us to see more clearly the situated nature of the historical materials we mapped in Chapter 1.

The US historical garden literatures are Europe-facing. They pay little attention to the agricultural knowledges and practices of First Nations peoples or of the nations in the South Americas. This omission is indicative of attitudes at the time. But Indigenous communities not only had important botanic, horticultural and agricultural knowledges and practices (often integrated with cosmological understandings) but also robust pedagogical strategies for their intergenerational transmission. Rather than recognise and use them, school gardens had a part to play in the erosion of such Indigenous knowledges, as Kay Whitehead (2018) notes in the Australian context. Her study of the pre-Federation school garden movement in South Australia (Pat's home state) shows that planting English trees, flowers, fruits and vegetables in school grounds was part of the colonial practice of place-making – reproducing and producing a Northern 'home' in the global South. When schools sent cottage garden cuttings home with students they materially displaced Indigenous connections with land. A little later, the Australian school garden realised a different white-nationalist imaginary – post-Federation, children were encouraged to plant school gardens in the shape of the map of a now constitutionally unified country (Holmes, Martin, & Mirmohamadi, 2008).

The nation-building role of school gardens can also be seen in Mateja Ribaric's (2017) account of school gardens in Slovenia where, in 1851 teachers were encouraged to teach fruit farming as part of an attempt to refocus the nation's agricultural production. But as the officially endorsed practices of 'intelligent agriculture' shifted over time so too did the school garden curriculum. By the post-war period, school gardens were a site for teaching children about the importance and practices of cooperative labour in line with contemporary socialist government thinking.

These are not the only examples of the 'moral imperatives' (Robin, 2001) of school gardens giving way to national economic and social demands, be they pro or anti the existing nation state. We mentioned in Chapter 1 Froebel's resistance to Prussian rule manifest in curriculum and in the school garden (Herington, 1998). But similar examples exist in other locations. Finola O'Kane (2000) argues that the garden of Patrick Pearse's boarding school in Dublin materialised Celtic Revivalism, the social and political struggle against Anglo-British culture and political dominance. Mao Zedong and Mahatma Gandhi advocated school gardens as educational practices which valued labour and valorised those who laboured. However, in both China and India the pressure for basic education and a vocational education which supported industrialisation held sway. A national school garden movement was not a feature of either of the national educational systems under their influence (Zachariah & Hoffman, 1985). School gardens, as Trelstad (1997) suggests, appear to be most successful when they are congruent with wider educational and/or national goals.

Of course, it is not just gardens and educators in the past who are situated in time and place. The work of current educational researchers is also socially situated. It is not surprising therefore that some of today's educational historians have:

- focussed on the formation of subjects and subjectivities. The historical materials we surveyed in Chapter 1 are now understood as positioning the garden as a site of moral learning and for performing gendered relations (e.g. Forrest & Imgram, 2003).
- documented the ways in which school gardens were integral to progressive pedagogies and radical forms of schooling with innovative formal curriculum and social learning (e.g. Burke & Dudek, 2010).
- examined gardens and food, particularly the connections between poverty, school dinners, allotments and school gardens (see, for example, Burke, 2005). While a food supply and school garden link was explicit in wartime (Hayden-Smith, 2014), nutrition seems not to have been a major concern for early C20 school garden advocates – rather the health issue was fresh air and exercise. It is in studies of boarding schools that we see a nexus of garden, eating and self-sufficiency (Edmundsen, 2009).
- interrogated assumptions about the natural world made in historical materials. The human-centric assumptions of garden-based learning about nature are readily apparent in the anthropocentric purposes attributed to school gardens (Gaylie, 2009). The walled kitchen garden is one obvious example of taming the wildness in human interests.

Our reading of contemporary historical research on school gardens suggests some future directions. There is already published material about public and landed-elite private school engagement with UK wartime gardens (Mayall & Morrow, 2011). We might know more about social class and UK school gardens if the UK equivalents to the US materials we surveyed in Chapter 1 were digitised and analysed as an archive. We were pleased when we found a digitised book on gardening for children by the celebrated British garden designer Gertrude Jekyll (1908). Her descriptions of servants and playhouses big enough to home one or more poor families made us think about the ways in which the discourses of rural and urban and the virtues of hard work are coded by unspoken class assumptions (a perspective that today's health education researchers are variously addressing, see next section and Chapter 5).

We have taken from contemporary garden histories the importance of wider social contexts for the introduction of school gardens, particularly:

- the role of the nation state, its place in the world and the construction of its imaginary via gardens;
- the significance of the philosophical and political orientations of garden advocates;
- the workings of class, race and gender; and
- the role of individual schools in deciding whether to establish school gardens.

Today's educational historians are not the only researchers to be concerned with wider social structures and/or discourses; this focus can also be seen in some of the current research on food and education, health education and environmental education.

Food, health and the environment in education

Researchers working on aspects of education related to food, health and the environment have focussed on the structures, cultures and discourses that shape what can be done in educational settings. Adopting more critical approaches, these researchers show how food experiences, health education, and/or environment programmes are shaped by forces beyond the school itself, as well as by school practices (Earl, 2018). Rather than exploring the benefits of eating more fruits and vegetables, for example, these (post)critical researchers look at the ways children are able (or not) to eat fruits and vegetables in the first place.

Much of this (post)critical research is based on Foucauldian understandings of power/knowledge. Researchers explore 'what else' is happening in schools, beyond the well-intentioned, 'doing good to others' (Guthman, 2008) of health–food–environment policies designed to teach children variously about healthy eating, encourage fruit and vegetable consumption, improve food and health knowledge and develop 'an appreciation for good food' (Waters, 2008). Issues of race, class and gender, as well as concerns regarding structures of inequality and the food system frame many of these investigations. As Flowers and Swan (2012) propose, the racialised, classed and gendered moralities of food knowledge are fundamental concerns for scholars, and the struggles over this knowledge are about 'legitimacy of authority' (p. 537). In other words, research on food and health in schools is as much to do with power as it is to do with fruits and vegetables.

In a special issue of the *Australian Journal of Environmental Education*, Swan and Flowers (2015) argue that there is a need to rethink food pedagogies beyond the narrow, 'gendered, racialised, and classed discourses of risk, obesity, healthism, and "gastronomification"' (p. 147). They go on to question the types of knowledge being created, taken up and used within food and sustainability education, and the knowledges that are absent – local, experiential, community, Indigenous, embodied, emotional, feminist. They prod us to ask, what kinds of 'truths' do we know about the world if these knowledges are obscured or silent? How might our understandings of 'problems' be different if these knowledges were consulted?

The critical, in food and health educational research, is key to moving beyond narrow, healthist approaches. (Post)critical scholars have highlighted what happens in schools when health and/or food policies are universally applied. Wright and Dean (2007), for example, examined how the 'truths' and 'imperatives' of the obesity epidemic are taken up and used in school textbooks and health information, without critical examination of the ideas being taught. Leahy and Wright (2016) showed that pedagogies and their intended effects are never straightforward, and may encourage problematic practices around food (like purging or starving). Leahy and Wright argue that bio-pedagogies strip away 'aesthetic, social, political and cultural complexities and entanglements' of food in everyday life (p. 243). The current pedagogical assemblages of health education are focussed on the individual and their responsibility for their own health, rather than on the structural problems that demand individual responsibility (Leahy, 2014).

Simplifying complex food practices to distil a singular message is a concern for researchers (e.g. Earl, 2018, 2020; Guthman, 2011; Hayes-Conroy, 2014). The tendency for policies to create easy, understandable, 'sound bite' advice regarding food and health – 'eating well reduces your chances of falling ill with cancer, heart disease, a stroke, or diabetes' (Dimbleby & Vincent, 2013, p. 30), for example – obscures the complexity of the food system, and individual food practices. If eating well to prevent ill-health was easy, surely we would all be doing it? Soundbite advice and policy hide the way the food industry keeps processed foods cheap. It also silences the realities of food poverty, diverse opportunities for/and practices of home cooking, and ignores different cultural understandings of health and well-being. Such simple policies then shape the everyday experiences of food in schools.

The issue of class and school food becomes particularly apparent through the lunchbox, where mothering is 'made visible' through the foods chosen (Harman & Cappellini, 2015; Morrison, 1996). Researchers in England (e.g. Parsons, 2016; Wills et al., 2011) have identified the way healthy family food practices are key aspects of middle-class motherhood and identity. In Canada, Cairns et al. (2013) show that the phenomenon of the 'organic child' is connected to symbols of middle-class motherhood. Investigating school food rules and policies in Australia and the UK, Pike and Leahy (2012) argue that lunchbox guidelines about 'good' and 'bad' food work as a form of surveillance that make some mothers come to be seen as 'defective', and moral judgements around good/bad mothering become normal. Such moral judgements can have unintended negative effects on children and their mothers (Tanner et al., 2019).

The issue of moral judgment is a serious concern for (post)critical scholars, because it is based in classed, racialised and gendered understandings of health and well-being. Those who do not fit the preferred norm are seen as defective and morally wanting (Pike & Kelly, 2014), and may be harmed by the resulting scorn and punishment (Vander Schee 2009). Health policy is not neutral, and those who fail to comply with rules or expectations are vulnerable to stigmatisation (Vander Schee 2009); such affects and effects seem to rarely be considered in health promotion campaign development.

Food, and the practices and foodways of individuals, are neither universal nor neutral nor simple. They are highly complex, shaped by structures, cultures and policies, histories and traditions. Engaging with the wider context brings a wealth of knowledges about ordinary people's lives to light, providing a much richer and varied foodscape than that focussed on 'healthism'.

Research on contemporary school gardens

Unlike historical and contemporary research on food, health and environment, most research on school gardens is not concerned with structures, cultures and discourses, but with questions about processes and outcomes. In education, this type of research is often characterised as a 'what works' approach (Davies, Nutley, & Smith, 2000). The focus is largely because governments want to see public services such as schools 'deliver' (Barber, Moffit, & Kihn, 2010) and funders of programmes want to ensure that their programmes do what is intended. But these are not the only reasons for the emphasis on outcomes. Schools are also interested in outcomes – they too want to see that their efforts have a student learning pay-off (Brown, 2015).

But in concentrating on 'what works', schools and teachers are now positioned to see a practice, intervention, event or treatment to be of value primarily if it can be shown to be so; this is often called performativity (Ball, 2003). The practices of performativity, being seen to be worthy, relies on making results visible, through the use of proxy indicators, categories and numbers (Loxley, 2007) – test results equate to learning, satisfaction survey scores equate to successful schooling. The press for tangible demonstrations of benefits and value lead to particular scientised forms of research (Lather, 2004). There is therefore in education, as elsewhere, a hierarchy of 'good' research with Randomised Control Trials (RCTs), case control studies, cohort studies and mixed methods studies at scale preferred to smaller, in-depth and often bespoke qualitative inquiries (Gough, Oliver, & Thomas, 2017). It is also widely held that policymakers are more persuaded by numbers than any other form of evidence, although public policy researchers point out that evidence is only one of a number of influences on political decision-making (Cairney & Oliver, 2018).

A 'what works' approach to evaluating and researching school gardens is thus not surprising, given that they often start out as an innovative project that the school takes on in addition to its usual curriculum. But school gardens do not readily lend themselves to RCTs, as there are too many variables to control for. There are, however, studies that use case control method, particularly in evaluations of garden programmes (reported later in this section).

We found only a few systematic reviews of school garden research. One, examining health and well-being (Ohly et al., 2016), noted the poor quality of quantitative garden research, saying that the evidence produced for changes in students' intakes of fruit and vegetables was based on self-reporting, and thus limited. By contrast, the qualitative research was better designed and thus more persuasive in its attribution of a range of health and well-being impacts, particularly for less academically successful students. The qualitative studies were process-oriented and thus better able to pinpoint barriers to success – for example, lack of funding and over-reliance on volunteers, difficulties in integrating gardening into the school curriculum. Another systematic review focussed directly on processes, examining research into teaching that promoted healthy eating (Dudley, Cotton, & Peralta, 2015). The reviewers examined the results of studies which looked at eight different teaching approaches, including the experiential (school gardens and cooking/eating). They concluded from their meta-analysis of 1 RCT, 13 quasi-experimental studies and 35 cluster controlled trials that experiential learning was associated with the largest effect on reduced food consumption, increased fruit and vegetable consumption or preference, and increased nutritional knowledge (c.f. Davis, Spaniol, & Somerset, 2015) – a result which would have pleased the philosophers and garden advocates from gardens past. Huelskamp's (2018) systematic review focussed on challenges to school garden programmes. In addition to finding gaps and deficiencies in published research, the review identified common recommendations – 'building a broad network of support in the community, providing professional development to teachers involved with the garden, and providing teachers with standards-based curricula to integrate the garden into multiple content areas' (p. 11).

But, rather than conduct another systematic review, we focussed our reading on (1) evaluations of school garden programmes, and then (2) the broad field of school garden research. We address each of these, concluding with our continuing heuristic for our own school garden research and analysis.

School garden programme evaluations and research

School garden programmes typically have a high-profile sponsor and take a (garden) fork-to-(cutlery) fork approach. They promote gardening alongside health and nutrition learning, cooking and links with the general school curriculum. School garden programmes typically offer external support and professional development as well as resources and links with other schools. A 'foodie' discourse overlaid with concerns about obesity and health underpinned all of the programmes we discuss (see Chapter 5).

The **Edible Schoolyard** (ESY) programme has operated in the US since 1995. Beginning in one middle school in Berkeley California, with the support of Chez Panisse chef Alice Waters, it now has a nearly 6,000 member programme across the country, according to its website[1]. A one-acre garden in Berkeley operates as a demonstration site and 'innovation hub'. The ESY programme is funded primarily by foundations and donations, and offers subsidised professional development programmes and network support to member schools and staff. A range of model lesson plans and learning resources, freely available on the website, are designed to support schools to get involved in garden-centred learning. The website also offers a bibliography of research into the value of school gardens, presumably as reassurance to those wanting evidence.

The ESY's progressive philosophical underpinnings are obvious on the website – 'designing hands-on educational experiences', 'placing the child at the centre of their education' and 'a just and joyful learning experience for every child'. Like its historical US school garden antecedents, ESY could easily

> claim philosophical roots in—and pose new philosophical questions for—now largely neglected traditions of educational thought on childrearing, coeducation, ethics of nourishment, educational aesthetics, ecological education, and schooling of impoverished communities.
>
> (Laird, 2013, p. 12)

Modern garden progressivism also embraces environmental concerns and ESY has been seen as a site where sustainable principles and 'eco-literacies' might be learnt (Murphy, 2003).

Nevertheless, research on ESY has food and nutrition as its primary focus. ESY evaluators (Rauzon, Wang, Studer, & Crawford, 2010) compared the food habits of fourth- and fifth-grade middle-school students in Californian schools with highly and less well-developed School Lunch Initiative (SLI) programmes, in which ESY is a key partner (SLI typically includes healthy school lunches, a school garden and classes in cooking). The evaluators found that SLI had a positive effect on children's eating habits both at home and at school. Children were not only more knowledgeable about food but also made connections between healthy eating and environmental concerns. The healing properties of ESY school gardens are also apparent in Fakharzadeh's (2015) account of post-Katrina New Orleans, where cultural traditions and local knowledges informed the ESY programme and this, in turn, had a positive impact on the surrounding communities.

The same positive conclusions about food education and eating habits appear in serial evaluations of the Australian **Stephanie Alexander Kitchen Garden** (SAKG). SAKG was begun by another celebrity chef interested in supporting plot-to-plate school programmes. SAKG offers 'pleasurable food education' with strong links to various subjects in the curriculum. Like ESY, SAKG offers schools membership of a network of 1,000-plus schools, newsletters and resources, telephone support and online and face-to-face professional development which can also, for a price, be onsite and bespoke to one school. Training in garden and cooking classes are offered to community volunteers, teachers and teaching assistants. The website offers 'proof that it works' in the form of three formal evaluations undertaken over several years[2] (see Figure 2.1)

The 2011–2012 national evaluation stated that the sustainability of the SAKG programme was dependent on the full commitment of the school principal and staff, ongoing collection of impact data, support for volunteers, managed staff turnover in schools as well as shared planning time for SAKG specialists and classroom teachers.

Figure 2.1 **SAKG evaluation results.**

Evaluation	*'What works'*	*Challenges*
2007–2009 12 primary schools in Victoria (Block et al., 2009)	• strong evidence of positive changes in children's attitudes, knowledge, skills and confidence in both gardening and cooking • no significant statistical difference on reports of levels of enjoyment from program and comparison schools • the higher the qualifications and experience of the garden specialist the more children reported enjoying their experiences. • new links between school and community	• lack of influence on home food practices • difficulties of recruiting enough qualified volunteers to run classes • ongoing expense of the programme • ensuring ongoing funding • integrating SAKG into the curriculum

2011–2012 representative sample of 28 SAKG schools and 14 comparators (Yeatman et al., 2013)	• significant changes in kitchen lifestyle behaviours, positive changes in students attitudes to healthy food choices, positive impact on students' social behaviour and skills • enhanced inclusion through sharing cultural food traditions, hands-on learning and preparing affordable meals that could be cooked at home.	• no difference in gardening behaviours • no statistically significant differences in eating habits • challenged by a crowded curriculum and lack of planning and teaching time • lack of adaptability to local needs • curriculum integration issues, • educational expertise of volunteers • working conditions and remuneration of programme staff • lack of project management, procurement, contractual and regulatory expertise
2018 Longitudinal element – 118 young adults in SAKG 2017–2009 study plus comparators (Block et al., 2019)	• cooking skills (76%); cooking behaviours (65%); enjoyment of school (59%); gardening (52%); food choices (52%); health (46%); wellbeing (32%); study choices (13%); and career aspirations (11%) (p. 5).	• not statistically significant when compared with non-programme participants.

In England, the **Royal Horticultural Society** (RHS) also promotes school gardening, signing up to one campaign some 11,500 primary schools. The RHS campaign provided resources but also a focus, structure and legitimation for gardens. Evaluators looked at 10 exemplar case study schools concluding that benefits included:

> greater scientific knowledge and understanding; enhanced literacy and numeracy, including the use of a wider vocabulary and greater oracy skills; increased awareness of the seasons and understanding

> of food production; increased confidence, resilience and self-esteem; development of physical skills, including fine motor skills; development of a sense of responsibility; a positive attitude to healthy food choices; positive behaviour; improvements in emotional well-being; and development of active citizens as well as independent learners and had observed changes not only in the children, but in attitudes to the school within the local community.
> (Passy, Morris, & Reed, 2010, p. ii)

The evaluators noted that, if they are to succeed, school gardens need: the active support of the head teacher and a key member of staff who drives the work in the garden, the amount of work needs to be manageable, and the garden needs a high profile within the school.

These are of course not the only school garden programmes and evaluations that are available. However, even across these three programmes there are some patterns, agreements and questions that arise. All of the programmes claim improvements in children's food and garden knowledge. All make strong claims for social learning gains and the positive effects of practical engagement for young people who do not fare well with bookish pedagogies. There appear to be variable connections with the curriculum outside of the more obvious health and personal development, and Science. The realisation of the multiple potential links into learning are jeopardised by the crowded curriculum/teacher workload. While external support for the programme is welcome, voluntary labour and school leadership commitment, knowledge and costs also play a significant part in the success and sustainability of garden initiatives.

The field of school garden research

We were struck by the relative paucity of critical work in current wider school gardens literature. Most researchers in the field report empirical studies but do not ground their analysis in educational and/or social theory, as do many historians, food/health researchers and ESD researchers. There is far more psychology and health-associated conceptual framing on offer. Because our own interest is in problematisations and their consequences, we were particularly keen to source papers to help us develop this line of inquiry, so in the following discussion we highlight papers that take a (post)critical perspective. We also include some wider literatures pertinent to school garden programmes.

We divide school garden research into two broad categories: (a) school change and (b) curriculum and teachers. There are of course multiple ways to think about a field of research and any division is somewhat arbitrary. And there cannot be hard boundaries between these categories, as many studies overlap and cover more than one topic. However, our goal was to guide the research and interpretations that follow in Chapters 3–6, mindful of potential blind spots.

School Change

'What works' garden evaluations (above) document a number of key school change issues. A 'consensus of expert garden practitioners' (Diaz, Warner, Webb, & Barry, 2019) summarised these as a shortage of time, limited resources and support; inadequate staff skills and qualifications; and challenges with education (class size, performance-based pedagogy) and the school curricula – difficulties in identifying links to educational objectives and standards and lack of curricular support. Low-income schools experience further difficulties working around the imposts of standardised testing (Gree, Rainville, Knausenberger, & Sandolod, 2019).

We take another perspective. We understand school gardens as an instance of 'vernacular change' (Hall & Thomson, 2017); where specific programme initiatives are always mediated by both state/national policies and local histories, populations, needs and assets (Loftus, Spaulding, Steffen, Kopsell, & Nnakwe, 2017). As Schafft and colleagues (Schafft, Hinrichs, & Bloom, 2010) put it, farm-to-school programmes are a 'flexible range of locally embedded strategies that schools might use to address specific community and school needs' (p. 23).

However, school garden literatures rarely address school change, leadership and management per se, and are most often disconnected from that field of study. For instance, in her book on the history of ESY, Alice Waters' (2008) narrative could be analysed to elucidate the importance of being part of a system-wide change (SLI); aligning external support with school district and school goals; the equitable allocation and everyday organisation of time, space, people and money; decision-making and governance; meaning-making practices; professional learning; and curriculum and assessment. However, it is up to the reader to do this kind of work.

By contrast, there *is* a literature on change and the leadership and administration of ecologically sustainable schools. The 'what works' research highlights administrative barriers to change – for instance, Kadji-Beltran, Zachariou and Stevenson (2013) suggest that some principals lack the administrative know-how necessary for developing sustainable schools and many are also unwilling to challenge the status quo. Principals need, they argue, professional development in 'empowering staff, encouraging critique of current approaches and exploring alternative possibilities for curriculum, pedagogy and policy' (p. 303). But other ESD researchers are more interested in change theory. Change theory is essentially a form of school improvement, where a school committee is formed, conducts a sustainability audit, develops an action plan including work on the curriculum, and monitors and evaluates and informs and involves the wider community (see for instance Eco-Schools Seven Steps https://www.ecoschools.global/seven-steps).

'Green school leadership' literatures often bring together sustainability science, research on the built learning environment and mainstream educational leadership/change (Kensler & Uline, 2016). Curriculum design, pedagogy and learning outcomes are connected with strategic visioning and planning, resource allocation, use of technologies, play and ground management, system support, legal frameworks and policy (Chan, Saunders, & Lashley, 2015). Green leadership literatures stress the importance of forming new professional norms and values (Leo & Wickenberg, 2013) and the daily iteration of the environmental mission through the words and actions of senior leadership teams (Ackley & Begley, 2010). Familiar barriers also appear in the leadership and change literatures – limited money, time, information and personnel being key to materialising enthusiasm for change (Veronese & Kensler, 2013). Perhaps surprisingly, we found little linkage of critical leadership literatures and school-as-an-organisational-ecology theory with eco-school or ESD change practices – Barlow and Stone (2011) are a notable exception. Nor do the ESD change literatures generally draw on the rich vein of theoretical work around sustainability, place and human–non-human interactions. We suspect that both ESD and garden research might benefit from engaging with school change literatures (but see Gough, Tsee and Lang 2020).

Most of the garden evaluation literatures mention volunteers, parents and the community. While there is some specific research into this group, most of it reiterates the benefits and the challenges of supply and management (c.f. Henryks, 2011). One important perspective on community participation, an ethnographic study of mothers who attempted to change school food in a school district, showed a less than homogenous 'good'. Stapleton (2020) demonstrates that 'through long-term persistence, the women created a path for progress that ultimately led to major change in the district's school food, yet their key role in laying groundwork for the change remained largely unrecognized'. While the dogged expertise of the mothers produced change, the gendered politics of recognition meant that their agency and contribution were glossed over.

School gardens are part of the built school environment and there is a small but important body of research which examines school yards. There is some evidence of the benefits of green school grounds; they are associated with positive changes in physical, mental and emotional well-being (e.g. Bell & Dyment, 2008). But researchers are divided in their view of how important the grounds are to school change – whether changing school grounds can lead change by raising awareness and changing attitudes (Izadpanahi, Elkadi, & Tucker, 2017), or are better understood as an integral and material part of a holistic school change process (Cole & Altenburger, 2019; Woolner, Thomas, & Tiplady, 2018).

Teachers and curriculum

Evaluators and researchers typically note the success of gardening programmes in teaching about food and nutrition, and about horticulture (Diaz et al., 2019). There are claims made for learning in other formal subject areas such as Science, Maths, English and Art and Design. SAKG evaluators (Yeatman et al., 2013) noted that schools generally linked gardening with Science, Technology and Maths, while kitchen activities were linked with English, Maths, Health and Physical Education. There are also strong claims made for cross-curriculum areas such as ESD and environmental literacies (Christodolou & Korfiatis, 2019; Hunter et al., 2020; Stone, 2009). Positive gains in social and emotional learning and a sense of well-being are also attributed to school garden programmes (Dyg & Wistoft, 2018; Roberts, Hinds, & Camic, 2019). We note that the historical tension we identified between the control of nature and a conservation ethic is rarely highlighted in these investigations (but see Gaylie, 2009 for one exception).

Teachers are clearly key to any change process in schools. They are frequently mentioned in garden evaluations but the research tends to see them as a conduit to benefits for students (as above). We did locate some research on teachers themselves. The finding that teachers who garden in their own time can be good mentors to their colleagues, while also promoting the value of gardening (Kincy, Fuhrman, Navarro, & Knauft, 2016), seemed unsurprising to us. We could not help but think of innovation literatures that suggest some early enthusiasts are necessary but not sufficient for successful implementation (Rogers, 1962/2003). We were more interested in research which suggested that teachers engaged in agroecology programmes in Mexico were highly likely to change their own eating habits and to value more highly local knowledges; but their knowledge of scientific process and agroecological principles remained disappointingly limited (Ferguson, Morales, Chung, & Nigh, 2019). This lack of disciplinary knowledge potentially undermined scaling-up the programme.

Teacher knowledges and repertoires of pedagogical practice are crucial to any garden programme implementation. Not all schools and teachers find it easy to manage the garden as a transdisciplinary (usually early childhood and primary) or interdisciplinary (usually secondary) curriculum, building knowledge and skills systematically, year after year (a challenge not confined to garden programmes, see Thomson, Hall, & Jones, 2012). An integrated curriculum which uses experiential pedagogies can easily lead to weakened disciplinary knowledge (Bernstein, 2000; Kneen, Breeze, Davies-Barnes, John, & Thayer, 2020); it takes considerable pedagogical expertise to plan and animate a curriculum organised around a specific activity, event or site.

The garden research literatures evidence the challenges of curriculum integration. Not all schools and teachers find it easy to make strong and ongoing connections with the garden and it can remain an isolated project or 'bolt-on' elective activity. Teacher attitudes are important in initiating and sustaining garden programmes. Some teachers are passionate about gardening and are able to pass that love on to students (Wistoft, 2013). Others may find the hands-on and open-ended pedagogies of garden programmes a way to resist performative policy agendas (c.f. Kemp, 2019). However, some are just unwilling to take up their spades and hoes and may make the 'whole school' goals of gardening programmes difficult to achieve (Passy, 2014).

Teacher reluctance is not the only issue at stake. Teachers' practices are always shaped by national policies and school interpretations (Ball, Maguire, & Braun, 2012), as well as by particular school populations and local assets and histories (Thomson, 2000). Thus, the ways in which teachers accommodate local and neighbourhood 'funds of knowledge' (Gonzales, Moll, & Amanti, 2005) vary, according to garden researchers. Rossi and Kirk (2020) examined SAKG and argue that when gardens are tied to national curriculum standards then the professional and local expertise of volunteers and communities is easily lost. This example stands in contrast to instances where school gardens have afforded opportunities to value rather than revile home food practices (Kiddle, Mawer, McAuliffe Bickerton, Ryan, & Siemonek, 2019); institute intergenerational learning (Mayer-Smith, Bartosh, & Peterat, 2007); recognise and use Indigenous knowledges (Narayan, Birdsall, & Lee, 2019), heritage languages (Mangual Figueroa, Baquedano-López, & Leyva-Cutler, 2014) and critical perspectives on race and immigration (Adams & Hyde, 2018); and engage with the politics of global food production systems (Mann, 2018) and the ethical choices it presents (Kiddle et al., 2019).

School garden programmes may present teachers with difficult practical as well as curriculum choices. Do they prioritise participation over the nurturing of plants (Hipkiss, Windsor, & Sanders, 2019)? How much agency is allowed to children? Do teachers insist on students being engaged in all aspects of gardening, or do they employ labour-saving devices to do the heavy-lifting and professionals the design, while children plant, weed and water (Mannion, 2003)? Is there inevitably a choice of sustaining student interest or the garden (Almers, Per, & Kjellstrom, 2018)? Can teachers incorporate the sensory and messiness of experiential garden learning (Surman & Hamilton, 2019) in contemporary modes of assessment? Does gardening inevitably prioritise the human over the non-human through its focus on cultivation and control (Mycock, 2019)? Might school gardens become a place for learning about environmental justice (Martusewicz, Edmundson, & Lupinacci, 2015)? Might the garden be both a site of resistance and hope (Bowers, 2005)?

Towards a heuristic for school gardens

Our historical and contemporary reading of the literatures suggest that it is important to understand school gardening as shaped discursively through at least four layers:

1. The wider social context – the production and reproduction of particular classed, raced, gendered and abled relations, practices and subjectivities: a globalised world economy and food production and distribution systems; globalised media and celebrity cultures; temporal events including war, climate collapse and food insecurity
2. The nation state and its public policy and education policy regimes including audit, innovation and improvement practices
3. The whole school – in its local context, its administration and ethos, improvement strategies and professional learning communities, its permeability to parents and wider community
4. The enacted formal and informal curriculum – its overarching philosophy and organisation into subjects, cross-curriculum emphases and classroom organisation

These layers are not separate, not nested Matryoshka-like, not separable laminations, but are imbricated like a ripple cake – interconnected through discursive flows of information, resources, people, technologies and truths. We think of this ecologically, as a complex assemblage where change happens in both predictable and unpredictable ways.

Our readings suggest that school garden research rarely puts this ecology together. We have attempted to do just this in the remainder of the book. In doing so, we do not offer a neat unitary story but hope instead to generate new insights, lines of inquiry and practical possibilities. The third layer, changing the whole school, is, as we have argued, a particular gap in the garden literatures and the two portraits and our concluding chapter particularly focusses on this layer.

Notes

1 https://edibleschoolyard.org/about
2 https://www.kitchengardenfoundation.org.au/content/pleasurable-food-education

References

Ackley, C., & Begley, P. (2010). The changing face of purpose-driven school administration: Green school leadership. In A. H. Normore (Ed.), *Global perspectives on educational leadership reform: The development and preparation of leaders of learning and learners of leadership* (pp. 377–396). New York: Emerald.

Adams, J., & Hyde, W. (2018). The critically designed garden. *The International Journal of Art and Design Education, 37*(3), 345–353.

Almers, E., Per, A., & Kjellstrom, S. (2018). Why forest gardening for children? Swedish forest garden educators' ideas, purposes and experiences. *The Journal of Environmental Education, 49*, 242–259.

Ball, S. (2003). The teacher's soul and the terrors of performativity. *Journal of Education Policy, 18*(2), 215–228.

Ball, S., Maguire, M., & Braun, A. (2012). *How schools do policy. Policy enactments in secondary schools.* London: Routledge.

Barber, M., Moffit, A., & Kihn, P. (2010). *Deliverology 101: A field guide for educational leaders.* Thousand Oaks, CA: Corwin.

Barlow, Z., & Stone, M. K. (2011). Living systems and leadership: Cultivating conditions for institutional change. *Journal of Sustainability Education, 2*(March), 1–29.

Bell, A. C., & Dyment, J. (2008). Grounds for health: The intersection of green school grounds and health-promoting schools. *Environmental Education Research, 14*(1), 72–90.

Bernstein, B. (2000). *Pedagogy, symbolic control and identity* (2nd ed.). London: Rowman & Littlefield.

Block, K., Carpenter, L., Young, D., Hayman, G., Staiger, P., & Gibbs, L. (2019). *What's cooking? Evaluation of the long-term impacts of the Stephanie Alexander Kitchen Garden Program.* https://www.kitchengardenfoundation.org.au/sites/default/files/Files/UoM_SAKG_Eval_Report_Final%20AUG%20 2019.pdf: University of Melbourne: Centre for Health Equity, Melbourne School of Population and Global Health.

Block, K., Johnson, B., Lisa, G., Staiger, P., Townsend, M., Macfarlane, S., Gold, L., Kulas, J., O'Koumunne, C.C., Waters, E. (2009). *Evaluation of the Stephanie Alexander Kitchen GardenProgram, Final Report.* https://mspgh.unimelb.edu.au/__data/assets/pdf_file/0010/2076733/SAKGP_Final_Evaluation.pdf: Deakin Univerity, University of Melbourne.

Bowers, C. A. (2005). *The false promises of constructivist theories of learning. A global and ecological critique.* New York: Peter Lang.

Brown, C. (Ed.) (2015). *Leading the use of research and evidence in schools.* London: Institute of Education Press.

Burke, C. (2005). Contested desires: The edible landscape of school. *Paedagogica Historica, 41*(4 & 5), 571–587.

Burke, C., & Dudek, M. (2010). Experiences of learning within a twentieth-century radical experiment in education: Prestolee School, 1919–1952. *Oxford Review of Education, 36*(2), 203–218.

Cairney, P., & Oliver, K. (2018). How should academics engage in policymaking to achieve impact? *Political Studies Review, 18*(2), 228–244.

Cairns, K., Johnston, J., & MacKendrick, N. (2013). Feeding the 'organic child': Mothering through ethical consumption. *Journal of Consumer Culture, 13*(2), 97–118. doi:10.1177/1469540513480162

Chan, T. C., Saunders, R., & Lashley, L. (Eds.). (2015). *Green school leadership: What does it really mean?* Pennsylvania, PA: IGI Global.

Christodolou, A., & Korfiatis, K. (2019). Children's interest in school garden projects, environmental motivation and intention to act: A case study from a primary school of Cyprus. *Applied Environmental Education & Communication, 18*(1), 2–12.

Cole, L. B., & Altenburger, E. (2019). Framing the teaching green building: Environmental education through multiple channels in the school environment. *Environmental Education Research, 25*(11), 1654–1673.

Davies, H. T. O., Nutley, S. M., & Smith, P. C. (Eds.). (2000). *What works? Evidence-based policy and practice in public services.* Bristol: Policy Press.

Davis, J. N., Spaniol, M. R., & Somerset, S. (2015). Sustenance and sustainability: maximising the impact of school garden health outcomes. *Sustainability and Public Health Nutrition, 18*(3), 2358–2367.

Diaz, J. M., Warner, L. A., Webb, S., & Barry, D. (2019). Obstacles for school garden program success: Expert consensus to inform policy and practice *Applied Environmental Education & Communication., 18*(3), 195–206.

Dimbleby, H., & Vincent, J. (2013). *The school food plan.* Retrieved from http://www.schoolfoodplan.com/wp-content/uploads/2013/07/School_Food_Plan_2013.pdf

Dudley, D. A., Cotton, W. G., & Peralta, L. R. (2015). Teaching approaches and strategies that promote healthy eating in primary school children: a systematic review and meta-analysis. *International Journal of Behavioral Nutrition and Physical Activity, 12*, 28.

Dyg, P. M., & Wistoft, K. (2018). Wellbeing in school gardens – the case of the gardens for bellies food and environmental education program. *Environmental Education Research, 24*(8), 1177–1191.

Earl, L. (2018). Schools and food education in the 21st century. Abingdon: Routledge.

Earl, L. (2020). 'Who likes fish? And I don't mean fish fingers!' Taste education, culinary capital, and distinction in a rural middle-class primary school. *Cambridge Journal of Education, 50*(1), 77–93.

Edmundsen, J. (2009). The Ackworth school garden and history. *Garden History, 37*(2), 226–230.

Fakharzadeh, S. (2015). Food for thought. The intersection of gardens, education and community at Edible Schoolyard New Orleans. *Children, Youth and Environments, 25*(3), 175–187.

Ferguson, B. G., Morales, H., Chung, K., & Nigh, R. (2019). Scaling out agroecology from the school garden: The importance of culture, food, and place. *Agroecology and Sustainable Food Systems, 43*(7–8), 724–743.

Flowers, R., & Swan, E. (2012). Pedagogies of doing good: Problematisations, authorities, technologies and teleologies in food activism. *Australian Journal of Adult Learning, 52*(3), 532–572.

Forrest, M., & Imgram, V. (2003). School gardens in Ireland 1901–24. *Garden History, 31*(1), 80–94.

Gaylie, V. (2009). *The learning garden. Ecology, teaching and transformation.* Dordrecht: Peter Lang.

Gonzales, N., Moll, L., & Amanti, C. (2005). *Funds of knowledge.* Mahwah, NJ: Lawrence Erlbaum.

Gough, D., Oliver, S., & Thomas, J. (2017). *An introducton to systematic reviews* (2nd ed.). London: Sage.

Gree, A., Rainville, K., Knausenberger, A., & Sandolod, C. (2019). Opportunities for school garden-based health education in a lower-income diverse, urban school district. *American Journal of Health Education, 50*(4), 257–266.

Gough, A., Chi-Kin Lee, J., & Tsang, E-P K. Eds. (2020) *Green schools globally. Stories of impact on education for sustainable development.* Dordrecht: Springer.

Guthman, J. (2008). Bringing good food to others: investigating the subjects of alternative food practice. *Cultural Geographies, 15*, 431–447.

Guthman, J. (2011). *Weighing In: Obesity, food justice, and the limits of capitalism.* Berkeley: University of California Press.

Hall, C., & Thomson, P. (2017). *Inspiring school change. Transforming education through the creative arts.* London: Routledge.

Harman, V., & Cappellini, B. (2015). Mothers on display: Lunchboxes, social class and moral accountability. *Sociology, 49*(4), 764–781.

Hayden-Smith, R. (2014). *Sowing the seeds of victory. American gardening programs of World War 1.* Jefferson, NC: McFarland & Company, Inc.

Hayes-Conroy, J. (2014). *Savoring alternative food: School gardens, healthy eating and viceral difference*. London: Routledge.
Henryks, J. (2011). Changing the menu. Rediscovering ingredients for a successful volunteer experience in school kitchen gardens. *Local Environment, 16*(6), 569–583.
Herington, S. (1998). The garden in Froebel's kindergarten: Beyond the metaphor. *Studies in the History of Gardens & Designed Landscapes, 18*(4), 326–338.
Hipkiss, A. M., Windsor, S., & Sanders, D. (2019). The girl with the garden gloves: Researching the affordances of sensual materialities in the garden. *Ethnography and Education*, doi.org/10.1080/17457823.2019.1698309.
Holmes, K., Martin, S. K., & Mirmohamadi, K. (2008). *Reading the garden. The settlement of Australia*. Mebourne, AUS: Melbourne Unviersity Press.
Huelskamp, A. (2018). Enhancing the health of school garden programs and youth: A systematic review. *The Health Educator, 50*(1), 11–23.
Hunter, D., Monville-Oro, E., Burgos, B., Rogel, C. N., Calub, B., Gonsalves, J., & Lauridsen, N. (Eds.). (2020). *School gardens and healthy diets. Promoting biodiversity, food and sustainable nutrition*. New York: Routledge.
Izadpanahi, P., Elkadi, H., & Tucker, R. (2017). Greenhouse affect: The relationship between the sustainable design of schools and children's environmental attitudes. *Environmental Education Research, 23*(7), 901–918.
Jekyll, G. (1908). *Children and gardens*. London: Country Life Ltd.
Kadji-Beltran, C., Zachariou, A., & Stevenson, R. B. (2013). Leading sustainable schools: Exploring the role of primary school principals. *Environmental Education Research, 19*(3), 303–323.
Kemp, N. (2019). Views from the staffroom: Forest school in English primary schools. *Journal of Adventure Education and Outdoor Learning*, doi.org/10.1080/14729679.2019.1697712.
Kensler, L. A. W., & Uline, C. L. (2016). *Leadership for green schools*. New York: Routledge.
Kiddle, R., Mawer, C., McAuliffe Bickerton, C., Ryan, A., & Siemonek, L. (2019). E hoa ma, in ate or ate tangata (My friends this is the essence of life): Meal-making as a pedagogical tool for learning about food politics. *Policy Futures in Education, 17*(7), 805–820.
Kincy, N., Fuhrman, N. E., Navarro, M., & Knauft, D. (2016). Predicting teacher likelihood to use school gardens: A case study. *Applied Environmental Education & Communication, 15*(2), 138–149.
Kneen, J., Breeze, T., Davies-Barnes, S., John, V., & Thayer, E. (2020). Curriculum integration: The challenges for primary and secondary schools in developing a new curricula in the expressive arts. *The Curriculum Journal*, https://doi.org/10.1002/curj.34.
Laird, S. (2013). Bringing educational thought to public school lunch. Alice Waters and the Edible Schoolyard. *Journal of Thought*, Summer, 12–27.
Lather, P. (2004). Scientific research in education: A critical perspective. *British Educational Research Journal, 30*(6), 759–772.

Leahy, D. (2014). Assembling a health[y] subject: Risky and shameful pedagogies in health education. *Critical Public Health, 24*(2), 171–181.

Leahy, D., & Wright, J. (2016). Governing food choices: A critical analysis of school food pedagogies and young people's responses in contemporary times. *Cambridge Journal of Education, 46*(2), 233–246.

Leo, U., & Wickenberg, P. (2013). Professional norms in school leadership: Change efforts in implementation of education for sustainable development. *Journal of Educational Change, 14*, 403–422.

Loftus, L., Spaulding, A. D., Steffen, R., Kopsell, D., & Nnakwe, N. (2017). Determining barriers to use of edible school gardens in Illinois. *Journal of the American College of Nutrition, 36*(7), 507–513.

Loxley, J. (2007). *Performativity.* London: Routledge.

Mangual Figueroa, A., Baquedano-López, P., & Leyva-Cutler, B. (2014). La Cosecha/the harvest: Sustainable models of school-community engagement at a bilingual program. *Bilingual Research Journal, 37*(1), 43–63.

Mann, A. (2018). Education for food sovereignity as transformative practice. *Policy Futures in Education*, /doi.org/10.1177/1478210318816251.

Mannion, G. (2003). Young people's participation in school grounds developments: Creating a place for education that promotes young people's social inclusion. *International Journal of Inclusive Education, 7*(2), 175–192.

Martusewicz, R., Edmundson, J. & Lupinacci, J. (2015). *Ecojustice education. Toward diverse, democratic and sustainable communities* (2nd ed.). New York: Routledge.

Mayall, B., & Morrow, V. (2011). *You can help your country: English children's work during the Second World War.* London: Institute of Education.

Mayer-Smith, J., Bartosh, O., & Peterat, L. (2007). Teaming children and elders to grow food and environmental consciousness. *Applied Environmental Education & Communication, 6*(1), 77–85.

Morrison, M. (1996). Sharing food at home and school: Perspectives on commensality. *The Sociological Review, 44*(4), 648–674.

Murphy, J. M. (2003). *Education for sustainability: Findings from the evaluation study of The Edible Schoolyard.* Berkely, CA: Centre for Eco-literacy.

Mycock, K. (2019). Forest schools: moving towards an alternative pedagogical response to the Anthropocene? *Discourse.*

Narayan, E., Birdsall, S., & Lee, K. (2019). Developing a context- specific model for kitchen garden learning programmes. *Asia Pacific Journal of Teacher Education, 48*(2), 112–131.

Ohly, H., Gentry, S., Wigglesworth, R., Bethel, A., Lovell, R., & Garside, R. (2016). A systematic review of the health and well-being impacts of school gardening: Synthesis of quantitative and qualitative evidence. *BMC Public Health, 16*, 286.

O'Kane, F. (2000). Nurturing a revolution: Patrick Pearse's school garden at St Enda's. *Garden History, 28*(1), 73–78.

Parsons, J. M. (2016). When convenience is inconvenient: 'healthy' family foodways and the persistent intersectionalities of gender and class. *Journal of Gender Studies, 25*(4), 382–397.

Passy, R. (2014). School gardens: Teaching and learning outside the front door. *Education 3–13, 42*(1), 23–38.

Passy, R., Morris, M., & Reed, F. (2010). *Impact of school gardening on learning final report submitted to the Royal Horticultural Society*. Slough: NFER.

Pike, J., & Kelly, P. (2014). *The moral geographies of children, young people and food: Beyond Jamie's School Dinners*. London: Palgrave Macmillan.

Pike, J., & Leahy, D. (2012). School food and the pedagogies of parenting. *Australian Journal of Adult Learning, 52*(3), 434–459.

Rauzon, S., Wang, M., Studer, N., & Crawford, P. (2010). *Changing students' knowledge, attitudes and behaviour in relation to food. An evaluation of the school lunch initiative*. http://www.schoollunchinitiative.org/downloads/sli_eval_full_report_2010.pdf: Dr Robert C and Veronica Atkins Centre for Weight and Health, University of California, Berkeley, CA.

Ribaric, M. (2017). *Nature's classroom – The school garden, yesterday, today and tomorrow*. http://museums.eu/article/details/121590/natures-classroom-the-school-garden-yesterday-today-and-tomorrow#accept: Slovenian School Museum

Roberts, A., Hinds, J., & Camic, P. (2019). Nature activities and well being in children and young people: A systematic literature review. *Journal of Adventure and Outdoor Education*.

Robin, L. (2001). School gardens and beyond: Progressive conservation, moral imperatives and the local landscape. *Studies in the History of Gardens & Designed Landscapes, 21*(2), 87–92.

Rogers, E. M. (1962/2003). *Diffusion of innovation* (5th ed.). New York: Free Press.

Rossi, T., & Kirk, D. (2020). The pedagogisation of health knowledge and outsourcing of curriculum development: The case of the Stephanie Alexander Kitchen Garden initiative. *Discourse*.

Schafft, K., Hinrichs, C. C., & Bloom, J. D. (2010). Pennsylvania farm-to-School programs and the articulation of local context. *Journal of Hunger & Environmental Nutrition, 5*(1), 23–40.

Stapleton, S. R. (2020). Nevertheless, they persisted: How a group of 'noisy moms' overcame dismissal and helped to improve school food in a US small city school district. *Gender, Place & Culture*.

Stone, M. K. (2009). *Smart by nature*. Healdsberg, CA: Watershed Media.

Surman, E., & Hamilton, L. (2019). Growing consumers through production and play: A phenomenological exploration of food growing in the school foodscape. *Sociology, 53*(3), 468–485.

Swan, E., & Flowers, R. (2015). Clearing up the table: Food pedagogies and environmental education - Contributions, challenges and future agendas. *Australian Journal of Environmental Education, 31*(1), 146–164.

Tanner, C., Maher, J. M., Leahy, D., Lindsay, J., Supski, S., & Wright, J. (2019). 'Sticky' foods: How school practices produce negative emotions for mothers and children. *Emotion, Space and Society, 33*, 100626–100626. doi:10.1016/j.emospa.2019.100626

Thomson, P. (2000). Like schools, educational disadvantage and 'thisness'. *Australian Educational Researcher, 27*(3), 151–166.

Thomson, P., Hall, C., & Jones, K. (2012). Creativity and cross-curriculum strategies in England: Tales of doing, forgetting and not knowing. *International Journal of Educational Research, 55*, 6–15.

Trelstad, B. (1997). Little machines in their gardens: A history of gardens in America 1891–1920. *Landscape Journal, 16*(2), 161–173.

Vander Schee, C. (2009). Fruit, vegetables, fatness, and Foucault: governing students and their families through school health policy. *Journal of Education Policy, 24*(5), 557–574.

Veronese, D. P., & Kensler, L. A. W. (2013). School leaders, sustainability, and green school practices: An elicitation study using the Theory of Planned Behavior. *Journal of Sustainability Education, 4*. http://www.jsedimensions.org/wordpress/wp-content/uploads/2013/02/LisaKenslerWinter20131.pdf

Waters, A. (2008). *Edible Schoolyard*. San Francisco, CA: Chronicle Books.

Whitehead, K. (2018). James Greenleas' school garden and the suburban dream in colonial Australia. *Studies in the History of Gardens & Designed Landscapes, 38*(4), 342–352.

Wills, W., Backett-Milburn, K., Roberts, M. L., & Lawton, J. (2011). The framing of social class distinctions through family food and eating practices. *Sociological Review, 59*(4), 725–740.

Wistoft, K. (2013). The desire to learn as a kind of love: Gardening, cooking, and passion in outdoor education. *Journal of Adventure Education & Outdoor Learning, 13*(2), 125–141.

Woolner, P., Thomas, U., & Tiplady, L. (2018). Structural change form physical foundations: The role of the environment in enacting school change. *Journal of Educational Change, 19*, 223–242.

Wright, J., & Dean, R. (2007). A balancing act: problematising prescriptions about food and weight in school health texts. *Journal of Didactics and Education Policy, 16*(2), 75–94.

Yeatman, H., Quinsey, K., Dawber, J., Nielsen, W., Condon-Paoloni, D., Eckermann, S., Morris, D., Grootemaat, P., Fildes, D. (2013). *Stephanie Alexander Kitchen Garden national program evaluation: Final report.* https://www.kitchengardenfoundation.org.au/sites/default/files/food%20education/sakgnp_evaluation_uow_finalreport_2012.pdf; Centre for Health Service Development, Australian Health Services Research Institute, University of Wollongong.

Zachariah, M., & Hoffman, A. (1985). Ghandi and Mao on manual labour in school: A retrospective analysis. *International Review of Education, 31*(3), 265–282.

3 City School establishes a garden

Why does a school become interested in having a garden? What do they hope it will do for their children? This chapter focusses on the way that one school established a garden, and the motivations, challenges and details of doing so. In examining the various literatures on school gardens in Chapter 2, we identified that school gardening is shaped discursively through multiple layers. This chapter and the one that follows provide portraits of two school gardens in different stages of development. We draw attention in these chapters to the whole school – discussing not only the garden itself but also school leadership, management, ethos, teachers and school staff, and permeability with community – and the way these different layers create a complex assemblage that shapes school change.

This chapter is a portrait of City School, a primary school in a gentrifying area of a large city in England. The portrait is divided into two parts. The first focusses on the purposes and processes of building the physical school garden. The second section examines the learning associated with the curriculum and children's experiences with gardening.

Purposes and processes

City School is located in a mixed suburb of a large urban city in England. The neighbourhood, traditionally working class, is gentrifying. Middle-class families (forced out of more traditional middle-class areas by rising house prices), immigrant families and working-class families all jostle together in the streets surrounding the school. The school itself is a large, red-brick, three-storey Victorian building that houses primary-aged children. There is also a nursery, located in a separate block. The school has large grounds which are mainly concreted, with the exception of the garden space on the right-hand side that can be seen by looking up at the school from the road. Protected all around by black iron palisade fencing, entry to the school is by intercom at a pedestrian gate. The streets surrounding the school are populated with large, leafy trees, some of which provide shade over the garden.

Inside, the main school building is a rabbit warren of corridors leading to secret study rooms, larger neutral spaces leading into classrooms and winding staircases. Classrooms have large windows and high ceilings; they are bright spaces. Outside, the new garden is developing. At the time of our visits, in April and June 2016, a greenhouse had been established on a grassed area. Large 'yeti' feet led from the asphalt playground to the greenhouse door. Behind the greenhouse was a bamboo forest, and to the right, the palisade fence and some flower beds.

The asphalt nursery play area is separated from the greenhouse by a short fence. The fence comes directly out from the nursery building creating a protected space, ensuring that the nursery children are not overwhelmed by the older children during breaktimes. The fence is also a musical instrument, with young children able to make notes and sounds by moving mobile, hanging pieces along the fence. Older children (Years Three to Six), have their own large playground set just above the nursery play area and the garden. The two playgrounds are separated by a gate. The older children's playground is also asphalt, with some fenced tennis courts where football is played over break. There are alleyways of nettles between the tennis courts and the fence. Footballs sometimes ended up amongst the nettle patch, so the nettles needed to be carefully negotiated; some Year Four pupils told us they disliked being stung by the nettles.

The garden space where the greenhouse sits had not always been accessible. There had once been a pond and an area used for outdoor learning, but these had fallen into disrepair and disuse over time, deemed dangerous, and so fenced off. Children had not had any grassed/natural spaces on the school grounds for many years. When investigations were begun into the feasibility of a school garden, much of the space was completely overgrown with waste material needing to be removed. The school manager painted the picture for us:

> at that time [of the school consultation with children] the garden was fenced off completely, was really uninspiring. Frankly dangerous: trip hazards, nettles, brambles, you name it, it was just a no-go zone and all the playgrounds were surrounded by what I call prison wire fencing, so 2.4 metre high chicken wire. Really

> unattractive, unappealing, not at all welcoming, not at all friendly. So hideous (interview).

The school staff were motivated to start the garden project by this legacy of wasted space, as well as a concern for children's resilience and exposure to nature.

Why a school garden?

At the time we visited, many of the children at the school lived in rented accommodation where there was no access to the outdoors. The neighbourhood was gentrifying, but 62% of children at City School spoke English as another language, and 38% of children came from disadvantaged backgrounds. Younger year groups (nursery and Year One predominately) had more children from privileged, middle-class backgrounds than the upper-year groups. According to the head teacher, 50% of children aged 7 and up were disadvantaged. One teacher explained how the changing demographic was made visible through gift-giving at the end of term:

> The Year One and Year Two children have 'powerful parents with good jobs'. Some of the parents are from The Times (media), from restaurants. You can see this through the gifts given to the teachers at Christmas – prosecco, candles, perfume, chocolates. Year Five and Year Six parents/children don't give presents to the teachers (notes).

In contrast to such displays of wealth, other families at the school experienced poverty and hunger on a regular basis. The school manager told us:

> There's quite a few children that would not have food in the morning so we offer breakfast club and support through Pupil Premium (a government subsidy for children from low income families) where required and the same with after-school club to make sure children do get fed. From time to time we do sandwich bags and things for parents to collect and take home, where they're in difficult circumstances and where they're accessing food banks and things like that, and maybe don't have access five days a week, seven days a week. We will help and they just go to the back of the kitchen and there's a little brown paper bag for them to take home and make sure that the children are coming to school fed,

> but also them and their siblings as well. I don't advertise it, it's just something we do on an as-required basis (interview).

The children talked about hunger in various ways in the focus groups we ran. For example, children in Year Three wanted to grow apples in the garden because then food would be available from the trees:

> I want our garden to have some tree that grows apples like if you're hungry…
> When you're hungry at lunchtime to just like oh yes… I'm gonna pick an apple. It's fresh so good, I didn't even rinse it. There might be worm inside. Yeah!
> They may add an apple tree because loads of people are hungry!
> LEXI: You think loads of people are hungry?
> Yeah but no I see they skip lunch and they don't know what to do.
> LEXI: Why did they skip lunch?
> Because… I don't know (Y3, FG1).

There were of course a myriad of reasons for a child skipping lunch, including not having brought food from home and/or being unable to afford the school meal provided. But children become aware and concerned about their peers' lives through the smallest of incidents and interactions.

An awareness of possible and actual deprivation framed the school's narrative about children's exposure to nature, and their knowledge of foods and food growing, and thus the potential purpose of the garden. Children were seen as lacking in opportunities to 'be' in green spaces – whether at home, at the park or, obviously, at school. As the head teacher explained:

> We firmly believe that being outside is really really beneficial to our children, many of whom live in flats and tower blocks near us so for us, we know how much it engages them to be outside and how empowering it is for them to feel that they can make a difference in the world in which they live, even from a very young age, by the choices they make, you know. Do we recycle or don't we recycle? Should we plant and grow or shouldn't we plant and grow? Should land be kept just for growing food or should we build another tower block on it? It allows for many many debates and discussions with the children which gives them the knowledge and the power to go forward fully informed of the effect they can have on their environment. And the footprint they'll leave behind (interview).

The head teacher saw that school gardens might move beyond a focus on food or health, to encompass a wide range of learning; gardening might afford a 'doing and being' that went beyond food education. The school garden problem was identified as children's lack of access to outdoor spaces, and with access came the ability to not only navigate the natural world but also benefit from it.

A lack of exposure to nature is often seen as detrimental to children's well-being and future health (see Louv, 2010). However, the driving nature deprivation narrative was not necessarily true for all the children at City School. The children we spoke to had varied experiences of green spaces. Some visited grandparents, who were farmers, during the holidays. Others relied on city parks. One young person told us that their grandmother had a greenhouse where they could grow tomatoes, while another mentioned their grandfather who grew tomatoes and cucumbers. These conversations illustrated taken-for-granted assumptions made by the school staff about children's exposure to gardens and food growing. We can only speculate about how many children saw food growing or gardens outside of school, but the existence of these different narratives suggests that young people were not as ignorant of the natural world as their school believed.

Constructing a new school garden: managing stakeholders

As with any project, the garden at City School had a variety of stakeholders, including parents, school management, external NGOs, teachers and children. The main stakeholders behind the garden concept – local parents, alongside the school management team – talked about the garden in terms of resilience. Resilience, they told us in various conversations and email exchanges, is found in the garden. And children needed to build their resilience to the world around them, in order to survive in it. Quite how they defined resilience was never made clear to us. Guthman (2008) and Pudup (2008) have argued that school gardening programmes in the US shape children into specific types of citizen-subjects – those who choose the 'right' foods, whose body shapes are 'correct' and whose expectations of food align with organic, local, authentic ideas of the alternative food movement. At City School, the ideal child was one who was resilient.

The 'ideal' child must have an ideal garden. City School's garden was designed by a celebrity gardener, in consultation with the young people and teachers of the school. The resulting garden plan was ambitious in scope and design, so much so that it was estimated by the school manager to cost upwards of £500,000. Obviously, the school did not have this kind of budget, nor did they want to develop such a project, and so the school manager and head teacher used the concept design as a basis from which to launch their own vision for the school garden. As the school manager explained to us:

> If the garden is going to be a blueprint for other schools to follow, and yes, an immaculate Kew Gardens type of garden would be something stunning for people to see but every school in the country is not going to be able to do that. They won't have the space and they will look at the cost and just say no. So I think actually, it is much more important to say, well, this is the concept, these are the ideas we wanted to incorporate, and this is how we're delivering it in a much simpler, much more cost effective way, and something that schools around the country could replicate (interview).

City School was keen to address issues of long-term sustainability and replicability. The garden at City School was funded initially through money raised via food festivals, auctions and donations (we will discuss these efforts later in this chapter). But, as the head teacher pointed out, such a strategy was not sustainable in the long term. Grants and other funding pots would need to be sought in order to make the garden sustainable. Dedicated funding from government is not available for school gardens in England, and so funds must be sought privately. The availability of private funds means that only some schools are able to afford gardens, and that many children miss out on gardening experiences and learning.

The multiple stakeholders involved in City School's garden development – parents, school staff, young people, and outside organisations – meant that maintaining ownership of the garden became a challenge in the long-term plan. As the school manager explained:

> That's actually probably one of the biggest challenges – in that there are so many people that feel that it's their project, that actually sometimes it almost becomes bigger than the school itself. And reigning it back is a challenge. But reign it back we will! At the end of the day, the garden has to be something that works for the school and for the local community. We do want the local community involved but it has got to be sustainable (interview).

The school manager's comments suggest some of the complexity of starting a school garden; one of her tasks had become managing the realpolitik of the garden project. While we were exploring the garden at City School, we came up against the many different ideas framing the work of the garden, something we will return to in Chapter 5, and experienced for ourselves some of the challenges the school manager encountered.

Some stakeholders saw the garden as primarily about food knowledge and growing. But the head teacher saw it differently; in words reminiscent of Montessori's concerns about Lucy Latter's horticultural approach, the head told us,

> It isn't just about growing and harvesting. It's actually about being outside and engaging with the environment and realising our part in it. So every three weeks every class will have been outside for either a morning or an afternoon.

The head saw that the inside and outside of the classroom had to be organised as educational spaces with diverse affordances for self- as well as teacher-directed learning.

> It's not just the garden we're developing. It's the whole outdoor learning area. In the summer we're spending a large part of our surplus on developing whole new play areas built out of logs for the children, so it won't just be the garden they have, they'll have so much more going on outside. The nursery is getting a mud kitchen which will allow them many more great learning experiences in the outdoors which is what we're trying to give them in a city setting. We're just very fortunate we have so much land and so much interest in doing so (interview).

The head teachers' comments illustrate how the vision for the City School garden evolved as the project developed – it moved away from a parent-led intervention towards becoming an integrated part of school life.

Raising the funds: festivals and foodies

In order to raise funds to get the initial garden project off the ground, City School undertook a number of fundraising endeavours, including an auction and a food festival. As the two parents spearheading the initial project were based in the hospitality industry, the food festival idea took root and grew. In the year we were researching at City School, they held a second food festival to raise further funds. One Saturday in June saw the playground transformed into a myriad of different stalls, all provided by leading restaurants in the city, each selling one small dish for a set fee. The stalls offered tiny tastes of some of the city's top restaurants and most trendy foods. Below are notes one of us wrote of the experiences of the day:

> When I arrive, it is raining but then not. In about five minutes the sun is back out and it is sweltering. The festival is packed with people already. There are several 'areas': the main playground is full of food stalls, there are two stages with performers located on the tennis court; in the area below, next to the garden (open for all visitors) herbs and plants are for sale (potted from the garden) and there is a Nando's truck. The school chef is running a barbeque on the edge of the garden. It is loud – there is a band playing and there are a lot of people.
>
> There are lots of teachers volunteering (later one tells me they had to do so) and they all seem in good spirits. I bump into the head teacher. She says, 'it's like Florida here today. They queue for Disneyland in the rain, so they queue here.' She is right. On the main playground food stalls are arranged around the edge, with communal tables in the middle. People can buy a plate and then sit and eat. There are tokens for sale – £2.50 for a purple token that buys you a small plate at one of the stalls, and £0.50 for a green token for the kiddies play area which is in the nursery playground.
>
> I overhear two people sitting on a bench opposite me, talking. 'Who would go for Nando's?' one asks the other. 'Unless it's easier for the kids?' The Nando's stall (not located with the other food stalls) does seem a bit of an anomaly amongst this sea of trendy, award-winning, fine-dining restaurants and their chefs. The chefs themselves are all here too – including famous names normally only seen on television. Dishes on offer include: duck liver eclairs; modern arctic rolls in bright colours; pork carnitas (pulled pork, fennel slaw and a brioche bun); dhansk veal mince with Bombay potatoes or puffed wild rice, and tamarind yoghurt; coq au vin brioche burger; coq au vin mac & cheese croquettes; deep-fried shark; meatballs; gelato; pizza; shrimp burgers; grilled mackerel fillets with piccalilli; ox hearts with green beans and tarragon; lamb cutlets; truffle mac & cheese; spiced lamb with flatbread; broad bean, pea, and ricotta bruschetta; fritto misto. People sip prosecco (£5 on tap) and eat from the small plates. Patrons seem to be people with families, younger people interested in food, other chefs. Queuing for food is necessary as there are so many people.
>
> The school chef grills chicken legs on the barbeque, served with salad and/or coleslaw. They have a fantastic spot, on the edge of

> the garden under a tree. He is in a good mood while I talk to him but their location (and that of Nando's) makes me wonder – is there a distinction between those stalls in the main playground (food by high-end chefs) and those in the lower playground (the school cook, Nando's)?
>
> Later there are performances by various groups from the school – the choir, the Portuguese club. The audience watching the stage (located on the tennis courts so slightly separate to the main food stalls) feels more representative of the school – mixtures of cultures and skin colours – whereas the food crowd is predominately white and middle-class (someone taking over The Guardian food Instagram for the weekend is here, taking photographs and posting stories). People queue for gelato and strawberries with beignets, strawberry Eton mess. By 3.30pm some stalls have sold out of food and are packing up. Families gather on the grass, relaxing. There are lots of activities, including cupcake decorating, face painting, hair braiding, balloon modelling, a Capoeira demonstration. It is an incredible event.

The food festival we attended raised £63,000 for the garden project (in six hours, as the head teacher was quick to tell us). As the head teacher and school manager explained, given the time and effort required of school staff to run the festival, this figure was not a 'true' one as it relied on so much volunteer time. As a result, the following year would focus on external streams of funding for ongoing garden growth. However, this was an extraordinary amount of money raised in a few hours.

But what does such a food festival reveal about City School?

The food festival provided a juxtaposition to the regular food environment surrounding City School. The streets around the school showcased the gentrification happening within the neighbourhood. Gastropubs jostled for space with Chicken Cottage and a local newsagent. Further up the street, a café made excellent toasted banana bread and flat whites. Many children attending the school were unlikely to have eaten at any of the restaurants featured on the main playground, but maybe had had Nando's (or similar). Bringing Michelin-starred chefs into a deprived borough speaks to a particular kind of 'food washing'.

The website for the festival sits separately to the school's main website, as if this particular kind of food event is not part of the wider school ethos or experience. A separate website creates its own narrative – one related to fancy chefs and posh restaurants rather than stories of the school. It speaks to a consumption-oriented lifestyle of which most parents and children of the school are not a part. And yet it is this festival that generates funds for the garden project in which all children can participate. There is a trade-off between the need for funds and the ways it is possible to generate such funds. And of course, the event raises questions for other school gardens – how do schools without access to networked parents or fancy restaurants raise funds for their gardens? If, as the school manager told us, the garden was to be a blueprint for other schools, how does this mode of fundraising detract from that goal?

The learning garden

The garden had only been 'officially' open since the spring. While it was being developed over the winter months, and at our first visit, the area was fenced off and restricted. By the time of our second visit, anyone was able to visit the garden during the school day, in theory anyway, as children had to seek permission and/or be accompanied by an adult in order to visit the garden.

Planting and gardening were overseen by an employee of a gardening organisation, hired to come in two days a week, keep the garden growing and also run sessions with young people. Once a week the teacher-gardener ran a 'gardening club' for Year One children after their school day ended. She was also present during the lunchtimes before these sessions so that young people in other year groups were able to participate in gardening work, and she hosted groups of children from classes when teachers were planning. Her teaching occurred largely in the greenhouse which was built while other work on the larger garden was going on. The greenhouse was quite small. In the centre were raised tables where plants grew in pots. All around the edge of the greenhouse were raised beds with plants. There was a tap for gathering water.

Children were introduced slowly to the garden as a new learning space, with opportunities to participate in willow-weaving classes, gardening club and other classes (such as art) where the teachers moved outside.

The school ran two training days focussed on using the garden – the first aimed at teachers, the second at teaching assistants. The training days, delivered by outside NGOs and other organisations, were designed to familiarise the teachers with the space, and show how the space could be integrated into the curriculum across all the year groups. Training specifically focussed on learning in Science and Art, Mathematics and English. The head teacher was keen to emphasise the importance of the garden not just being a 'token' space – it must be fully integrated into the curriculum and wider school life. Training days were important not only because they helped staff invest in the garden concept but also because it showed staff members how the garden might be integrated into their teaching plans.

Edible gardens and the curriculum

> I eat the vegetables that grow in the garden and lots of them are like leaves and there's radish in all of them but I think they're trying to grow carrots now (Y4).

The importance of 'knowing where your food comes from' is a phrase shaping edible education across the world. As societies have become less connected to agriculture and farmers, so alternative food activists have sought to link ordinary people with the growing that produces food. Alice Waters, original promoter of slow, local, homegrown foods in the US (see ESY Chapter 2), speaks romantically of school gardens as

> a way of making sure that children grow up feeling the soil with their own fingers, harvesting its bounty in the American sunshine, and watching their own hands make the kind of beautiful, inexpensive food that can nourish the body and the spirit.
>
> (Waters, 2008, p. 40)

The garden at City School navigated different opinions of what the garden might be for, as is illustrated by the school manager:

> The garden is really important because what we're hoping is that the children will have access [to] understand about food and understand about sustainability of food, to understand how to eat healthily, how to grow things, where food comes from. That's one part of it but it's a much more holistic approach because I really want to see the children have access to outdoor space to take physical activity. There's so many of the children just go home and play on their PlayStation or Xbox or whatever and actually love being out in the garden and the freedom to run around the garden and it's so obvious just in these last couple of days we've been back this term, where we turfed the garden and we've taken down the fence, that the children are loving being on the grass. That's a pleasure in itself. It's important that all children, regardless of their backgrounds, get the opportunity to play on grass and eat and understand food and where it comes from (interview).

But 'knowing where your food comes from', can be achieved using other approaches, as we learned through conversations with the lead gardener. For the gardener, the purpose of the garden was less to teach children about knowing where their food comes from and more about making links between what is growing in the garden at any particular moment, and the children's real lives. She explained:

> When I started this December I did some activities in class relating to food, making links between seeds that we eat on a daily basis, through plants, how we collect them, things like that. Now [in April] it's growing, growing, growing, growing. But then at the same time, I'm just showing the children what needs plants have to grow. The life cycle. We've already collected a couple of seeds so they were able to see that from one plant, which was kale, gave roughly 600 seeds. Then we've counted how many plants we can have next year from this years' seeds. It was absolutely amazing because you still eat the leaves, you still have the seeds for next time so that's all the life cycle. And I think it's a different usage of the plants I show them… That's actually bamboo, which grows over there, you can buy bamboo socks, you can do a house, you can make arts and crafts and stuff like that so it's not only a plant which you pass on a daily basis but it has a real use. I did ask them 'what sort of plants you had for breakfast?' And they said 'huh?' And I said 'well what did you have?' 'Oh I had Weetabix' and I say 'well, it's wheat and we pick the grass and then I have some seed.' For today I'm going to plant some wheat with them. Wheat is basic of so many foods. I just show them that you can do this and this and this and this and that from a single grass (interview).

The gardener's comments show potential formal connections to curriculum work on Science – life cycles, seeds, plant growth – as well as the potential for plant use in art classes, and connections to everyday eating. Conversations with children about their food choices (Weetabix) were not framed by the language of right/wrong food choice, or imbued with moral judgment, as often happens when young people talk about the 'processed' foods that they eat (see Earl, 2018; Pike and Kelly, 2014). Rather, connections were made to the plant origins of these foods (in this case, wheat), information that the gardener could then use to shape her plans of what to grow in the garden.

The gardener's approach broadened an edible garden beyond simply learning about which foods to eat. Rather, it began to make links to a wider food system, one beyond the simple growing-cooking-eating model often used in schools. It educated young people about the kinds of plants they were eating, and allowed diverse home and community experiences to be connected to the garden, including potentially, national curriculum learning. But what the gardener did outside did not appear to travel easily into the formal curriculum.

The garden at City School was intended to be integrated into the everyday curriculum, to become an extension of the bricks-and-mortar pedagogical space. Teachers, however, struggled to see how such integration might be possible, given the pressures they were already facing. The Year Six teacher explained:

> It [the garden] needs to be supporting the learning, not we need to spend three lessons on something else. With this curriculum becoming larger we've got certain demands on our time, especially as Year Six teachers, we've got so much we need to cover and that's now all through the school, there's so much content you can't be waste…I suppose the hesitation I have is that it can't be a wasted time. It can't be an excuse to waste time, especially in the mornings when we are focusing on literacy and maths. But we do have time in the afternoons where it certainly should be a space where people can be creative as well (interview).

The concern for the Year Six teacher, as for others, was that the garden did not necessarily lend itself to teaching the mandated curriculum. Even though aspects of Science might easily be taught through gardening, and indeed was being taught by the gardener, this was neither recognised or wanted by most of the teachers. Their understanding of the potential learning affordances of the garden did not extend to Nature Study which, as we discussed earlier, was historically connected to primary school Science. Teachers understood that trips to the garden had to be purposeful, and tied to the curriculum, but they were not yet clear how they would accomplish this.

The Art teacher did however see some connections between the curriculum and the garden:

> I have been trying to do some spatial drawing with the kids in the garden. I prepared them for it by having real plants in the classroom first and we'd do some careful looking and drawing and trying to get them to draw what they see. However, when we went to the garden they were just 'oh so yeah, we're in the garden' that the drawings were not great because they were just drawing something from their imagination. They weren't really focused. I think they

> couldn't focus because of everything else that was going on out there. So for me, quality learning and teaching is when they can take something quality that you may have done within the classroom, move it to an outdoor context and it brings something new to learning that you wouldn't have achieved within the classroom. It has to enrich what you're doing in the classroom and not take away from it. I think that's results really, isn't it? You want to see. But it's also their happiness. They still gain something from it, they've had a nice experience. That's great but have I got what I had planned to achieve in the end? Mmm not always [laughs]' (interview).

The Art teacher understood the value of scientific practices of observing, looking closely and drawing, as in botanical drawings and paintings. However, her standard classroom methods and discipline just did not work well outside, and her expectations about 'product' and productivity were unmet. Of course, teaching outside is not necessarily the same as teaching inside the classroom 'box'. The gardener offered one potential model of how to teach outside; the school might also have looked to Physical Education, forest school and Outdoor Education specialists who have expanded 'outside' teaching repertoires. But the Art teacher's initial negative garden experience was shared by her colleagues.

Experiences within the garden: control and freedom

In conversations with staff at City School, we discovered that learning to act 'appropriately' in the garden was one of their early concerns. Essentially, this meant that children needed to be supervised so that they did not damage or destroy parts of the garden, did not 'run wild', obeyed instructions, and generally acted in the same controlled manner as they did inside the classroom walls. Staff reasoned that because gardens are part of the wider school environment, the rules and regulations that governed other spaces also governed the garden. What goes in the classroom goes in the garden. This approach made the school garden different from, say, a local park, a garden at home, or a botanical garden where perhaps the 'rules' are less overt.

These disciplinary requirements were not meant to undermine any freedom the children might experience in the garden but to make sure everyone was 'safe' within the space, and nothing was damaged or destroyed during garden visits.

> I feel like our energy with the kids changes when we're out there, once they get over their initial excitement of just being outside.

> When they can start to see it as a classroom. They were even saying today, when we were in the greenhouse, like 'this is so peaceful' cause its really quiet once you're in there, especially when there's not too many kids. They are just excited because they get to go outside. But I've realised after the last couple of days that to then expect them to do focused work out there is sort of the next step because they just need, at the moment, to be exposed to being out in the space, understanding how to behave out there, without seeing it as a play area because they're used to only going outside to play (Art Teacher, interview).

Concerns about control, safety and behaving correctly are interesting, and perhaps contradictory, in the context of a space that is supposed to allow children the freedom to roam. There is, of course, a justifiable need for children to obey instructions, to not act in ways that damage the garden and to avoid unnecessary risk. However, narratives of control sit at odds with narratives of forest schools and garden spaces which focus on learning how to manage risks, having the freedom to explore and make mistakes, and becoming independent.

The need for control within the garden – both in the sense of behaving in a particular manner (i.e. not losing control in the green space) and in the sense of mastering one's own desires – links back to Victorian-era priorities for children. Cultivating children so that they know the limits of particular ways of being has been linked to the Victorian need to mould children into particular beings, and linked to the idea that gardens made people civilised (see Chapters 1 and 2). Indeed, the very idea of walled kitchen gardens is related to enclosure from the wild, keeping nature at bay and exercising control (Morris, 1996).

The notion of the garden as an educational space where one has to be controlled changes the nature of the experience for children. Those we spoke to thought of the garden as a calming space, a space for relaxation, where they could forget about things. But if children are being made to behave in particular ways, that freedom for simply *being* is lost.

The head teacher linked the need for children to learn to behave in the garden, an unstructured space, as important for children's long-term futures:

> If having an unstructured environment you feel leads to less learning then you really haven't embedded in your children a degree of independence or resilience that you want them to have a go at the world. If you feel you have to monitor within and can only

> keep them within the confines of the classroom to learn and to develop then you've failed them. They really will need that self-discipline and that resilience that allows them to thrive in any environment and take responsibility for their own behaviour and their own actions and ultimately their own learning. So, we feel that the more we have them outside in an unstructured dynamic environment, the more prepared they'll be for an ever-changing world (interview).

Resilience and independence here come to mean following the rules, self-managing within them, not requiring overt discipline from teachers. This is freedom to choose to behave in given ways, to follow the social order of the school (Hunter, 1994).

Children were aware that they needed permission to access the garden space, and could only be there under certain conditions. Teachers emphasised the need for 'control' within the garden, and how children needed to exercise self-discipline when they were outside. There is tension between this approach with the need for children to be exposed to natural spaces to learn about self-managing risk and becoming resilient. But this tension arose, we suggest, because the staff viewed the garden as a classroom, and therefore a space which needed the same set of school rules, rather than a 'natural' space with less rigid, different or no rules.

The Year Six teacher explained to us:

> I think it's about instigating that [behaviour] within the classroom before you then go out into a space. An outdoor space, you lose the confines of control I guess [laughs] in a way. You have to reset expectations and make children understand that as well, which we've had in PE as well. You go outside, this is how you behave. It's about setting those expectations but the only way that we're gonna be able to really get that across the school is using it repeatedly in every year group. Using outdoor spaces and using them constantly. If you have six months where you're not outside it gets to April now… they've spent six months not experiencing what the expectations are for outside. It's about becoming intrinsically involved in your lessons at all times (interview).

There were challenges for staff, trained largely to teach inside and to whole classes, in developing outside routines and practices which build self-care and care for others and the environment. There were implications for professional development and for partnership work with garden staff that might be taken up in the City School garden sustainability plan.

The garden as a place of soil, compost, vegetables, safety

The connections between 'natural' gardens and gardens-as-classrooms interested us. They became even more interesting as the idea of 'safety' emerged from conversations we had with teachers and children. We became interested in the garden as a 'place' that the children understood to be safe. They meant not just a space to feel safe, but one that was physically safe from nettles or glass or stones. The garden as a safe space was important to young people. There appeared to be two aspects that were key: the safety of plants within the garden and children's feelings of being safe within the garden. Concern for plant well-being came up in various focus groups we held with children.

> I like this photo because it has cages around it so it's healthy for the plant because they won't die because when people come past and they don't realise there's plants, they won't tread on them. The plants have a fence so they don't pass away (Y6, FG3).

Children wanted the plants growing in the garden to be safe from little feet, from being trodden or trampled on. These conversations illustrated a concern for the natural world, and demonstrated a care that did not fit with teacher talk of children running 'wild'. It also gave us pause to think of the arguments (such as those put forward by Louv, 2010) that young people's lack of interaction with the natural world will make them less inclined to care for it as they grow.

This care for plants in the garden seemed to also relate to young people's preferences for 'organised' spaces. In photograph elicitation sessions, children showed preferences for gardens that were neat, with plants in rows and grouped together.

> It's just really organised and neat. The plants are really organised.
> It's not messy. You'd have maybe daffodils in one row, purple flowers, it's organised.
> I think in a garden it's really good [to be organised].
> If someone needs something, they know where to go.
> And especially if it is symmetrical, it's really nice. I like things symmetrical, especially the garden (Y6, FG4).

Safety and organisation seemed to be interlinked. Navigating one's way through the garden was important, as was ensuring that plants thrived. Children's desires to feel safe within the space may at first appear to run at odds to adult wishes that they develop resilience and learn about risk, but perhaps these are related. In order to take risks, we must first feel safe.

Safety and organisation also related to soil and dirt. The children had strong views about what they disliked in the garden photographs we showed them. The compost heap, for example, was almost always disliked for being 'messy' or 'dirty', while the photographs of lavender, tomatoes or organised spaces were enjoyed. The photograph we used of a compost heap was strongly disliked in most of our focus groups, often being picked as the least favourite photograph. Reasons for this dislike varied, but often the compost heap was seen as a disorganised and unsafe space.

> It's not organised. It's not a garden. It's not anything (Y6, FG4).

Children were very concerned that the photograph showed wasted food, something that has become increasingly prevalent in the media and also relates to concerns around hunger.

> There is no point of growing food because they're taking it out from the roots and they're throwing it in the bin. They're wasting stuff (Y5, FG3).
>
> Look like in here, the fruit is all nice and fresh. All the fruit that is here is nice and fresh. Here it is not nice and fresh so they left it. The farmers don't really care about the farm. They do care about the farm but not as much because here is really nice and fresh and on [photograph] number 6, they left the food to rot and this is why they have so much waste. When it is ready you should pick it so you can sell it so you can get more money and you can eat it because it is good if you eat fruit (Y4, FG4).

Children were aware that food waste was not a good thing. Awareness of waste is fundamental to a conservation ethic, as is recycling and regeneration through practices such as composting.

Our research suggested that learning the basic principles of the life cycle might usefully help children build understandings about sustainability, and urban/rural life and their intersections. However, this would require the next step for the school to be engagement in serious curriculum development and planning.

Moving forward

City School was navigating the complexities of establishing a school garden with multiple stakeholders, agendas and aims. At the time of our research, the garden was in its infancy, and there were competing narratives of what the garden should do for the school, for the children and for the community. Everyone had different opinions, and it was not yet clear how the garden might be integrated into the school curriculum and thus used as an additional teaching tool on a daily or weekly basis. Such challenges are common to schools, and we will discuss how these different discourses shaped how the garden could be conceived in Chapter 5. In the chapter that follows, we provide a second portrait of a school with an established garden, one which had already done the planning and development yet to be undertaken at City School.

References

Earl, L. (2018). *Schools and food education in the 21st century.* Abingdon: Routledge.

Guthman, J. (2008). Bringing good food to others: Investigating the subjects of alternative food practice. *Cultural Geographies, 15*, 431–447.

Hunter, I. (1994). *Rethinking the school. Subjectivity, bureaucracy, criticism.* Sydney: Allen & Unwin.

Louv, R. (2010). *Last child in the woods.* New York: Atlantic Books.

Morris, M. S. (1996). 'Tha'lt be like a blush-rose when tha' grows up, my little lass'. English cultural and gendered identity in The Secret Garden. *Environment and Planning D: Society and Space, 14*, 59–78.

Pike, J., & Kelly, P. (2014). *The moral geographies of children, young people and food: Beyond Jamie's School Dinners.* London: Palgrave Macmillan.

Pudup, M.B. (2008) It takes a garden: Cultivating citizen-subjects in organized garden projects. *Geoforum,* 39(3), 1228–1240.

Waters, A. (2008). *Edible Schoolyard.* San Francisco, CA: Chronicle Books.

4 New School maintains an established garden

How do schools sustain their gardens? What can get in the way of an ongoing garden programme? What happens when the person who was most committed to the garden leaves?

In this chapter, we present a portrait of New School, our second case study. Our research at New School was conducted over two different time periods – in 2008 and again in 2012, as part of two separate research projects. In 2012, the head teacher interviewed in 2008 had moved on, and the school was in a period of transition under a new head. We have combined these two research experiences to create a narrative here about an established school garden and the way it survives (or not) a transition.

New School/eco school

New School primary is located in a deprived suburban estate, on the outskirts of a market town. The surrounding suburb is made up of post–World War II (1950s) council houses with front gardens, and the streets are quiet. The nearest chip shop and corner store are several minutes' walk away and there are no nearby supermarkets – you have to head into town for those. Of the 350 pupils on roll in 2012 when we revisited the school, 188 were eligible for free school meals (54%). The area is considered very socially depressed, scoring 54.25 in the Indices of Multiple Deprivation Combined Rank and ranking 1,516 of 32,482 areas, with many families out of work or living on benefits. The school is located on a stigmatised estate, and the school carries the legacy of the neighbourhood reputation.

The New School site is a mix of older buildings and a new eco building. The school is separated from the road by hedging that transforms into tall fencing and a pedestrian gate with a modern intercom system. A large biofuel tank stands near the school entrance, welcoming visitors to the eco-school. As you walk towards the reception, the single-storey dining hall is on the right, bordered by flower beds and fruit trees. Beyond the dining hall are further classrooms used for extracurricular activities, including cooking club and farm school. Then comes the farm where pigs and chickens are raised. Across the courtyard, a reception area leads into a large hall and then into the early years and nursery classrooms, older buildings whose hasty post-war construction creates ongoing maintenance problems. Opposite is the new two-storey 'eco-building' and beyond that a massive playing field. The eco-building has large floor-to-ceiling glass windows overlooking the playing field. The eco-building is the home of Years Three to Six classes, staffroom and library.

The eco-school dream belonged to the head who was at the school at the time of our first visit (Thomson, Day, Beales & Curtis, 2010). When he started in the mid-1990s, the school was in bad shape. Buildings due to be replaced in the 1960s were still in use, and largely not fit for purpose. The junior school and infants' school were run entirely separately. There were poor connections with the local community. As the head teacher explained:

> The school was painted a horrible, institutional green with brown carpets and the kids were really off the wall in more ways than one. A lot of the staff had been there a long time. It had had one of the first trial inspections – one of the trial Ofsted's – done by the local authority and it came out very badly on all counts and the behaviour of the children was really poor. The whole thing was drab and dreary and terrible really and that's why a couple of the candidates pulled out because they probably felt they couldn't hack this. But me, wanting a challenge at the time, being young and fresh and stupid probably, thought that if I couldn't make my mark here I couldn't make it anywhere so I saw it in terms of the only way is up and that's how we tackled it from the word go (interview).

The head teacher's vision was for a school that brought the primary and early years together, a new building, a sustainability focus, as well as respectful relationships with the community. A member of staff told us:

> [the head teacher] has always been a little bit of an eco-warrior but I think that ethos has just started to spread. And he's always had this vision of the school being the hub of the community and

> leading the community. And he's seen that sustainability in its complete form – sustainable life styles; sustainable relationships – he's seen that as the way forward and that's how it's evolved. I think he wanted to lead that in this area because it's an area that he thinks needs it. It's probably half and half really. He's always been an eco-warrior but he's made the link with sustainability and sustainable life styles and seen that perhaps the estate needed that leadership (interview).

The head teacher animated his eco-school vision through environmentally sustainable design linked to environmental education for children.

> I've always been very keen on the eco agenda because that's my training and also my background – countryside living and sustainable living. So that heavily influenced the way that I wanted to make the curriculum creative and relevant to the children and hands-on with lots of experiential learning. So environment education was a big part of that from the word go. When I became a head, I wanted that to become the hallmark of this school and I think the kids that we deal with find that particularly valuable anyway. So that was translated into the curriculum pre- and post amalgamation and it also translated itself into the desire to run sustainably, again, before it became particularly fashionable to do so (interview).

Such a perspective might seem at odds in our current climate-aware, sustainably focussed world, but in the late 1990s/early 2000s, sustainability was still a relatively radical idea. Rather than a myriad of different stakeholders with their own agendas, as was the case with City School, New School's changes were driven by the head teacher, the consequences of which we will discuss later in this chapter.

The eco-building became the hallmark of the intentions and ethos of the eco-school. Entering the new building, the first impressions are a feeling of space, light and a sense of solid structure. It is a timber-framed construction, built to exacting sustainability standards. The attractive structural beams are clearly visible. The amount of space in the classrooms is very impressive, much larger than in many new builds. On the ground floor there is also what is called the atrium, a large space running alongside the classrooms where children can spread out for creative play and group activities. The atrium has huge windows with very low broad sills/seats overlooking the field. This is a space for meetings, events and the library; there is also a highly visible meter displaying the amount of rainwater collected and how much has been used to flush toilets and so on. In addition to the doors joining classrooms, there are large partitions between rooms which slide back to create even more space for year groups to work together. The partitions are very substantial, really like wall sections, and have very good sound insulation, unlike many found in schools 20 or 30 years ago.

Two sets of stairs which come almost straight down into the atrium provide access for the children. There is a third stair for emergency use by the Year Six classes at one end of the building. There is also a lift for children with physical difficulties and for heavy loads. The stairs seem quite narrow but have an open structure providing a view of the atrium and the outside. At the top of the stairs is a slightly curving balcony, with an open railing, running the length of the building. This is another large open area accessible from classrooms offering areas for role play, group work and seating. The balcony not only offers an expanded learning space but also provides a very good view of the field. From the upstairs classroom windows, the views are of the Key Stage One roofs, alive with cress, and then the houses beyond, looking towards the centre of the town. From here it is possible to see, dominating the playground, a very large map of the British Isles. The map is part of the all-weather surface, made from recycled yoghurt pots by a local company – their contribution to the new build project.

The initiating eco-school head also oversaw the establishment of the farm and garden. Raised beds were built outside and each year group was encouraged to grow vegetables on their own plot. A pond and wildlife area were created to allow children to watch the natural world. A farm school taught children how to raise poultry and pigs, and later, how to pluck, gut and cook chickens. Opportunities for the community to participate in the school ranged from being able to buy eggs from the school farm chickens, to a parent's association, training classes and open encouragement to join children in the dining hall at lunchtimes. Things were different at New School.

In creating a learning community, the head teacher practised what he preached. He completed a doctorate studying sustainability. The approach which the head developed was deeply embedded throughout the school. It was however experiencing some strain by the time of our second visit.

Learning about farming and 'from field to plate'

The wailing and shrieking of excited pigs is an unusual sound on an ordinary school day but at New School, it was commonplace.

The school raised pigs to slaughter weight. Most afternoons of the week Foundation and Year One children were learning the art of animal care and farming in small groups of eight.

In addition to pigs, the school had four young chickens hatched on-site and a further half dozen large chickens which laid eggs. The pigs and chickens were kept in separate areas – the pigs' home was large and muddy, surrounded by a large green fence. The chickens were kept in an area opposite the pigs; they had a variety of coops and perches. This area was also home to the farm school classroom, with a newly built wooden shed in the centre. Children were able to hang their coats up inside, shelter from the weather, and store farm school materials – clipboards, wellingtons and jackets. The size of the shed restricted the size of the farm school class to around ten. During a regular afternoon session, children learnt about 'becoming a farmer' by first of all changing into coats and wellingtons (Earl, 2018). They also learnt the importance of hygiene and care for animals, and farm-to-plate eating and sustainability.

> At 2pm, I join farm school with eight Year One children. This class have been doing farm school for a year now so are becoming acquainted with the animals. Respect, and not being scared are important elements of the programme. We go out in a line to the farm school classroom which is in a different building from the Year One classroom, across a paved/asphalt open space. There is a square garden in the middle with fruit trees and rhubarb plants growing and a path through the middle, and there are flower beds with more fruit trees along the edge of the dining hall building. On the way over Ms B stops at the fruit trees in the square and asks what is growing on them. The class say: apples, plums, and pears. She then points out the rhubarb. 'What is that?' 'Pumpkin!' 'No, it's really good in crumble', she hints. 'Apple crumble!' someone shouts. 'No, actually it's rhubarb', she says. One of the girl's comments, 'We grow that in our garden at home'. Once in class, everyone sits on the carpet and they discuss the pigs and chickens. Ms. B talks about how the piglets were small when they arrived and now they're growing and eventually we will eat them. The class nod seriously. 'What can we eat of pigs?' she asks. 'Bacon,' one girl, answers. 'I love bacon,' she adds in a hushed voice. 'Ham,' Ms. B says. 'Who likes ham on their sandwiches?' The children nod. 'And pork roast too,' she adds. 'You take the skin off,' says the girl, 'and then you eat what's inside'. 'And the chickens give us eggs.' Ms. B tells the children (fieldnote).

The farm is integral to New School's sustainability programme and its eco-school status. Children we spoke to were aware of the gardens, and the kinds of foods grown within. Many had experienced Farm School classes in previous years, and some were part of the 'kill it, cook it, eat it' programme run by the farm teacher for children with additional learning needs, which we discuss below.

Food education at New School

The garden at New School was not a singular space but made up of a variety of different spaces and areas around the school, developed over the years. There were flower beds along the outside of buildings where vegetables (sprouts when we visited) were growing alongside poppies and fruit trees. The nursery classrooms had their own small outdoor garden that also housed a rabbit run. The large playing field had a herb garden to one side. There were raised beds. And along the side of the eco-building was a wildlife area complete with a pond, and there was cress growing on green roofs. Children told us,

> We've got a whole garden
> There's gardens for each year group
> We don't use them that much now though.
> We plant plants, we got composting. Basically, just like a normal garden, but a bigger garden.
> We grow apples, raspberries
> Carrots
> Strawberries, green beans, peas, potatoes (Y5, FG4).

The farm school teacher, Ms. Baker, explained her plot-to-plate philosophy:

> It's essential that children know where their food comes from, and how to source it within a natural environment. They need to be able to know about foraging skills that have been lost to us for generations. Food doesn't come out of a microwave or a tin, you go out and collect it, you gather it. It's seasonal. You have to wait for certain times of the year. I actually do run a small group where [we] have a food for free book and there are natural foraging areas in school that we go around and we help ourselves. I've got the children collecting the sweet chestnuts from the tree now. They know that they can collect them, they know that they can hand them in but they're not allowed to eat them. If parents come to

> me and say I would like my child to sample them, I will send some home in the parent's hand where they can be tasted at home with their parent under their responsibility. We have wild mint growing and I teach them that you can brush your teeth with it. We've got a Tudor knot garden that's a little bit over run at this time of year [November] cause it's just been all strummed right back, where there is rosemary, thyme, sage, lavender, cow parsley, mint, chives, a range of things. We call it a taste session so they can understand that some things taste nice in their own right and other things need to be cooked and other things are just to flavour (interview).

Ms. Baker saw the farm serving the same food education goals.

> We initially started off just with chickens and we ran for two years just to make sure we were covering all the legislation, to make sure that the children coped well with the fact that these were not pets, they were going to be eaten. They coped very well so we introduced pigs and we started off from the onset that these are not pets, these are part of the food chain, and they were fine with that. When they went to slaughter we waited three or four months before we introduced anymore, just to make sure parents, children and staff could get used to the idea. So far nothing negative has come back from the community, or the school. So that is why we keep running (interview).

The programme had, in fact, progressed from raising animals and sending them to slaughter, to involving children in the whole process from raising the animals, to dealing with death, to preparing the meat for cooking, to cooking and then, finally, eating it.

The school had become part of the Food for Life programme as a result of concerns about children's lack of understanding about the origins of their food as well as the food they were eating. Ms. Baker remembered:

> Before we bought any animals here, to get our Food for Life recognition, a lot of the children just thought everything came from the supermarket which shock, horror to someone like me. So we did a big survey and it just came to that we were seriously, seriously lacking in that area so we introduced the chickens. From the chickens I got lots and lots of numeracy lessons, data handling, counting in twos, threes and sixes because of the egg cartons. Lots of recycling because I was making the children responsible for bringing in the

egg cartons so we didn't have to go out and buy them. Then we were selling the eggs to the community and to the children's parents. The money that we raised we put into a little bank so we had a flow chart of what we'd made, who we'd sold it to, the children practically handled the money. I had more of my behavioural and special needs children willing to learn because it was a real life, first-hand experience. Then we did numerous cooking activities. We tried to cook eggs in many different ways, and compared them to cooking on hot stones in the sunshine to boiling them, frying them, microwaving them, scrambling them. Then we did the kill it, cook it, eat it session for the first time. I was very nervous about that to be quite truthful because I wasn't sure what feedback I was going to get. So I did it to a small target group of children that were under my personal care for education, called active learners. Quite a complicated letter went home explaining the process to parents, and they had to agree. There were three parts, they could take part in the plucking and gutting [the chickens were killed on site by an experienced chicken culler], then there was the cooking and the preparation of vegetables which were all grown on site and then there was the actual sitting down and eating of the meal. That way I thought by breaking it into three areas, at least everybody within my group would've been allowed. It went down a storm. There wasn't a single child that wasn't allowed to take part in any of it. Then all the scraps were saved including the carcasses, the vegetables, everything was saved. Then we put it all in a great big cook pot, and we made a soup and we used the herbs from the school garden to flavour it. And then we took an old, traditional style soup kitchen around with plastic cups for everyone just to have a little bit of a taste in it. So our first three chickens that were killed we had a meal for ten people and we made enough soup for 75% of the school to actually taste. The second time we did it, we slaughtered some of the pigs so I got some of the meat back and brought it into school and we cooked that alongside some of the chicken and we actually invited parents in for that meal. Eight out of the ten parents from my programme took part and were very, very happy to do so (interview).

The farm food education experience encompassed some basic skills teaching and also created a vehicle for improving school-parent and community relations. Not only were the community able to buy eggs from the school chickens, but parents could be involved in the kill it, cook it, eat it sessions through supporting their children and helping in preparation. The head teacher in 2008 made school–community relationships a priority, and saw a diverse and different curriculum as key to improving interaction and relationships between parents, the local community and the school.

Connections with the formal curriculum

Combining farm school within a garden programme under an eco-school umbrella shows the complexities in how school garden education is defined. Adding farm to garden means far more learning affordances – not only numeracy but also Science, literacy and the arts – observation, note-taking, close looking and the like. When children went on foraging trips with Ms. Baker, they were required to pay attention to the different plants around them, to listen to instructions and learn what was edible by looking for certain characteristics of a plant – the leaves, or the fruit, for example. Above all, exposure to plants and animals taught children about care for the natural world, something that came up in our conversations with children at City School too. Education with animals also built productive habits – persistence, patience and observant care – learning to recognise when the animals were happy, what their needs might be, paying attention to whether their food needed refreshing or their water needed changing, observing the mud and the puddles, looking for eggs in the hen coop, being outside in all weathers. Spending time outdoors allowed children to learn about seasonal change, and the way animals reacted to these differences. They were also connected to sustainability education about recycling and local eating.

The field note below illustrates how seasons, eating and recycling were all brought into a farm school class.

> All the chickens in here are quite big already so Ms. Baker is generous with the food. Each child throws two handfuls. They are then told to pick up some rubbish which can be thrown away. They do a little exploring and someone discovers that there are chicken eggs in the coop. 'Really?!' Ms. Baker says, in exaggerated surprise. 'Chickens don't lay eggs in winter'. She then says that they'll come back and take a look but first we must get rid of the rubbish. We all traipse in a line down to the bins and throw the rubbish into one of the big drums. Then we walk back up to the chicken coop. 'Now,' Ms. Baker begins, 'some children have found some treasure in the chicken house but I'm not sure because I don't think chickens lay eggs in winter.' The children are indignant. 'Yes they

> do!' 'We've seen the eggs!' 'Have you?' Ms. Baker asks. 'Well, let's go find out.' We let ourselves back into the coop and the children gather around the house entrance. Ms. Baker asks for volunteers to go in and have a look. Three children go in but it's dark so they can't see. She makes them all come out again and then they go in one at a time. Each comes out triumphantly with an egg. They all look excited. 'There's a baby chicken in here,' says one girl. 'No', Ms. Baker corrects her, 'those eggs are for eating. You can take it home and have it for your tea' (fieldnotes).

These are the kinds of life lessons that have been attributed to school gardens from Comenius onwards. Seeing and experiencing for yourself is a powerful beginning and motivation for formal classroom-based instruction. It can also enhance and exemplify concepts fundamental to further knowledge-building – the life cycle, as illustrated in the chick and egg, is a threshold concept in Science.

Sometimes, farm school allowed connections to wildlife, and highlighted the importance of biodiversity. For example, during one farm school class, Ms. Baker noticed a mouse in the chicken coop. One of the boys noticed it too. Everyone stood very quietly to encourage the mouse to reappear but it did not before we had to return to class. The farm school teacher was excited about this appearance because it meant wildlife was being attracted to the garden. It was a particularly interesting exchange, as the school cook had been complaining about mice the very same day, because builders were disturbing the foundations. Another time, a conversation about hedgehogs revealed the different ways people live with nature.

> T asks Ms. B where her hedgehogs are. (The hedgehogs were found on the farm earlier in the week and I'm also curious as to where they are being kept, as she seems to have hidden them away somewhere.) She says to him 'I'm not telling you T. Your grandfather eats hedgehogs and he'll come and steal mine if he knows where they are.' Curious, I say, 'your grandfather eats hedgehogs?'. Ms. B answers for T and says how his grandfather is a proper Traveller and keeps Traveller traditions like eating hedgehogs. She says it's why he is so fearless and comfortable outside with the animals (notes).

Exchanges about eating hedgehogs and standing quietly for mice contribute to an education in sustainability and eco-consciousness. Such experiences teach children respect for the natural world, and the way we might need to alter our own behaviour (for example, by standing still) in order to interact with (and look after) that world – an important lesson in the age of the climate crises. These experiences teach children to be curious and they provide opportunities for diverse cultural exchanges, such as the one above about Traveller traditions. Children are encouraged to discuss different food traditions, using their family and community funds of knowledge. Older children might pursue the tensions around food practices and conservation and the settlements that have been reached in other locations, for example, about the rights of traditional land owners to hunt and fish.

In short, the garden at New School provided opportunities for children to learn beyond the confines of the classroom about matters within the formal national curriculum.

Changing leadership, changing priorities

When the initiating eco-school head left in 2012, priorities and practices of sustainability had to make way for improving standards at New School. A difficult Ofsted report caused the new school leadership to reassess where their focus lay, and to set new priorities on standards in Mathematics and English. This was disappointing but not unexpected in the educational environment in England. Curriculum pressures, teacher time and emphasis on results have led to a narrowing of the types of experiences children can have at school. Ever more pressure is heaped on teachers to show progress; time to spend outdoors, away from formal classroom learning, has become very limited, particularly as children get older. Outdoor experiences are not as easily measured as those in the classroom, and so their value in the current English education system often goes unnoticed. Children's learning is shaped by a focus on SAT scores, progress in certain subjects and a narrow idea of what counts as knowledge (see Chapter 2).

However, the new New School leadership still saw sustainability (including gardening, farm school, food practices, and sustainability education) as a desirable feature of the school. As Mrs Perry, the deputy head teacher explained, the main aim of their food philosophy was

> For children to understand food and to appreciate food and you know, to value food. And, really just to explore food more than just, it's something you shovel in your mouth and don't think much about it. I'd like to encourage the children, as they grow up, to actually cook from scratch and things like that rather than rely on processed foods (interview).

Mrs Perry explained how things had changed post-Ofsted and change of head

> Each year group had a raised bed and he [the head teacher] insisted that they went to garden it. We had a rotation of different groups for the children to experience but that was the theory. In reality, teachers didn't…If they weren't interested in gardening then really, it was a bind. And it wasn't really done properly. So last year I said, the beds are there, I'd encourage you to use a bed and let your children experience gardening and growing, it's a fantastic experience for the children, [but] not going to force you. If you do want a bed, that's fine. If you don't, then Ms. B [farm school teacher] will cultivate those areas. We have had gardening as a dinner time activity. We did employ somebody who worked with small groups in the garden but they moved on and then you haven't got the people who've got that interest and enthusiasm. It's difficult to carry it on (interview).

Gardening activity decreased once the head teacher left the school. However, other forms of eco-school were strongly retained. On one of our visits, we noticed that the display outside the Year Five classroom had changed to reflect autumn.

> There's a tree branch with coloured leaves, felt spiders, and bats for Halloween, details on bonfire night and books about woodlands, scary stories, a book called 'Pumpkin Soup', and another called 'Witches Way Deadly Desserts'. There are fake pumpkins, pine cones, and a pomegranate (notes).

Displays like these speak to the school's wider eco-school philosophy as well as national commercialised events like Halloween. Such indoor displays survived the change in leadership and continued to connect young people with the natural world. Other connections to the seasons were made through events like the Harvest Festival assembly. Children learnt songs such as 'Easy Peasy Harvest', 'Cultivate', and 'We Plough the Fields'. These referred back to the local area's history as a farming community, where many children would originally have had farming backgrounds, and where a strong Traveller community with links to the land still existed.

Children told us

> In harvest season we do the harvest festival and we do how the harvest, for people who can't afford so we gonna, give tins and things and then we kind of give them to people that can't afford things. (Y5, FG1).

New School's children came from homes where parents often struggled to make ends meet, but they learnt at school that other people needed their help. They collected and donated long-shelf-life foods for those in need. Such activities speak to a wider social sustainability agenda, and make a connection with education for sustainable development.

Key elements of the eco-school philosophy remained despite the change in leadership, and influenced aspects of the school such as food procurement (see Earl, 2018). Under the new head teacher, much of the emphasis of outdoor learning lay in the farm school. As he explained:

> I think it's [farm school] powerful. It's good. In terms of our own school we've got as you've seen, pigs, chickens. That's our standing curriculum. And before I got the job here that was the way of the school, and it's still gonna be the way of the school cause what I'm adding to the mix is making sure standards are top priority as well. And getting the proper links between core subjects and excellent curriculum. So, we don't just go out to see the pigs because we've got pigs, we're actually tying this in with the curriculum. And you can tie it into science, you can tie it into English, you can tie it into all numerous things. But we've gotta get the quality links, not just going to see the pigs and chickens' (interview).

The issue of 'standards' dominated all ambitions for the school when we visited in 2012. The Ofsted judgment – that the school needed improving – meant a strong focus on results in Mathematics and English. As a consequence, non–core curricula activities had been sidelined or moved to after-school clubs. As a Year Five teacher explained to us:

TEACHER: It [the school] has changed hugely. The new head has come in and you know, the focus now is the basic skills, like literacy and numeracy because, let's face it, it was needed. We've all got our own plots for instance, we never get to them, I've not planted anything in my plot for two years now. It's just not the emphasis any more.

LEXI: Is that just curriculum, time pressure?

TEACHER: It is very much so, it is. And it's the threat of Ofsted coming and it's not, possibly, passing it and so, you just have to teach the basics well and unfortunately.... It was a trade-off. We didn't really struggle because the emphasis was not on the literacy and the maths.

LEXI: So it's the way the emphasis on things has shifted?

TEACHER: Absolutely. I mean they were still done but they weren't done to the degree that they're done now. You know as a teacher

> then you felt it was more important to be creative, to be outside, to [be] visiting the local farms, and talking about food, field to fork (interview).

As the priorities and focus of the school shifted, New School found itself in a period of transition and change, with an unknown and uncertain future. Ms. Baker, the farm school teacher, was still optimistic regarding the future of farm school, and outdoor learning more generally

> When the old head was here, masses of support. He was very similar to me. We both have very similar ideas and values. Obviously it's changed now. Mr X is in charge and he's well on board with it. He loves the idea. It looks like everything is going to move pretty much in the same direction with the same level of support but obviously being a new head in a new school, and we are a struggling school, we are a registered charity, we are in a difficult area, the main emphasis at the moment is on literacy, well Maths and English. That's why there's only the lower years that are on the rota for the farm group. But he assures me, once standards are up, we can take it on and just keep it rolling, and keep it going and growing. So I would say we have very good support from senior management. In a nutshell, I can't say too much about the current leadership team, can I? Because it's just changed but all early signs are fantastic and great and he buys the pork, he has the apple sauce, whatever we've produced he's shown a massive interest in. He's even wanting to source some chicks for his children for Christmas. So I mean whoa. That's really good early signs that is, isn't it? And we will be moving onto ducks soon. I've actually got some ducks waiting to go in there. So the ducks will manage it as well. I know that's not necessarily eating but we will eventually have duck eggs, which we'll sell and the children will taste but for me it's about the animals controlling the environment in a natural [way]. That's just as important for me as well (interview).

During our time visiting New School in 2012, we observed the garden and outdoor learning spaces being used mainly by children in younger-year groups. Older children's access to the outdoors was restricted to formal school breaks when they could run on the playing fields, and observe the pigs and chickens in their pens. The way children were exposed to the garden(s) at New School was changing, and priorities were shifting according to national educational agendas and requirements. We pick up these changes in the final chapter.

References

Earl, L. (2018). *Schools and food education in the 21st century.* Abingdon: Routledge.

Thomson, P., Day, C., Beales, W., & Curtis, B. (2010). *Change leadership. Twelve case studies.* NCSL commissioned report. School of Education: University of Nottingham.

5 Analysing the school garden

We have drawn portraits of two different schools' approaches to gardening to show how school gardens were established, maintained and evolved. This chapter steps back to explore the different ways that these two school gardens were constructed discursively; that is, how they were linked to wider societal contexts and problems.

Our examination of school garden histories and the survey of research literatures suggested that rather than addressing a singular problem, school gardens were seen as the solution to multiple and intersecting problems (Chapter 1). Each of our two schools conceived of 'problems' differently and drew on different discursive framings. In turn, these discourses shaped the way that the gardens were used, resourced, shared and talked about. These different problems were part of broader discursive assemblages in each school, with different emphases and, therefore, practices.

We identified four dominant garden discourses at work in our two research sites. The discourses (Table 5.1) we identified are healthy eating and foodieness; hunger; well-being and resilience linked to nature reconnection; and access in the garden.

In this chapter, we explore the way these four discourses shaped experiences, activities and trajectories. The chapter begins by examining the different discourses we identified in both schools. We will discuss each of these discourses in turn, and then examine the ways they were variously taken up by our schools, and by others. As will be seen, these garden-specific discourses were often combined with other educational discourses, for example about development – what children do when – and discipline – the need for children to do as they are told and develop self-discipline.

The importance of knowing where your food comes from

Healthy eating and foodieness were used at both our schools as a justification for school gardens. Foodieness is 'that discourse belonging

Table 5.1 Discourses of school gardens

Discourses of school gardens	
Healthy eating, foodieness, obesity	Gardens expose children to fruits and vegetables. This improves children's willingness to try new foods, and their acceptance of fruits and vegetables. A fork-to-plate approach helps children understand where their food comes from. This, in turn, helps children be healthier, ultimately avoiding becoming obese.
Hunger and poverty	Children are poor and do not get enough to eat and/or they eat the wrong foods. Gardens can help teach children about foods they may not encounter due to their life circumstances. In turn, children can help to teach their families about eating properly.
Well-being, resilience, sustainability	Exposure to nature is important for well-being and developing resilience. Resilience is seen as an individual trait which provides the capacity to manage the unpredictable, perhaps a 21st-century equivalent of the 'stiff upper lip' and 'Blitz mentality'. Teaching children to care for nature is important, given the ongoing ecological crisis and threat of climate change.
Access	Gardens are spaces for children to both play in and learn from. Children do not have opportunities to be in nature outside of school, particularly in cities. Connected to nostalgic and romantic ideas of the rural.

to foodies' (Earl, 2018, p. 4). By this we mean the ideas, talk and practices around food-purchasing decisions; knowledge about where food comes from; cooking and sharing meals; valuing food; producing food; and connecting food to politics, society and culture. In recent times, the concerns about food systems that have emerged are situated within foodie ideas of 'the production and consumption of food, the value we attach to food, and the importance of both cooking and eating in modern lives' (Earl, 2018, p. 5). Coupled with the focus on foodieness is an elevation of the importance of healthy eating amongst young people in order to avoid the serious threat of obesity.

Education policymakers see a lack of healthy eating – as advocated by nutrition science through the UK government–promoted Eatwell Plate for example – as a serious problem. In *The School Food Plan*, an English policy document focussed on improving school meals, the authors argue:

> This country faces a serious health crisis caused by bad diet. Almost 20% of children are obese by the time they leave primary school at 11. Diet-related illnesses are putting a huge strain on the

> nation's coffers – costing the NHS £10 billion every year. We need to tackle the problem now, before the costs (both personal and financial) become too heavy to bear.
>
> (Dimbleby & Vincent, 2013, p. 7)

The solution? Schools that weave food education (i.e. foodieness) – 'cooking, growing vegetables, even modest efforts at animal husbandry – into school life and the curriculum' (Dimbleby & Vincent, 2013, p. 7). Such schools must also provide healthy lunches and nutrition learning, food education being a signal to the wider agenda on improving child health. It is thus no surprise that improving children's eating habits, encouraging them to taste different foods and eat more healthily, and teaching them about *knowing* where their food comes from, were some of the aims of both New School and City School.

Healthy eating is an important narrative running through a lot of school gardening practices. Rather than being outdoor spaces for play or reflection, school gardens are educative – places where children can learn to eat healthily. They are, in effect, used as a health promotion tool (Beery, Adatia, Segantin, & Skaer, 2014). We can see this discourse enacted in other gardens besides City and New Schools, for instance in Alice Waters' Edible Schoolyard (Chapter 2) and Michelle Obama's White House Kitchen Garden in Washington, DC (linked to a wider programme of reform in US schools, including improved school lunches and the *Let's Move* campaign, all designed to address the problem of childhood obesity). But healthy eating also frames research undertaken on school gardens, as we explained in Chapter 2 (D'Abundo & Carden, 2008; Davis, Spaniol, & Somerset, 2015; c.f. Lineberger & Zajicek, 2000; Wells et al., 2018), which, in turn, shapes what we can know about school gardens and their place in schools and society.

At first sight, the garden seems logically connected to healthy eating. The garden is a place where fruits and vegetables are grown, potentially tasted in their raw state or otherwise cooked for tasting. But the connection to healthy eating is not quite so simple. The causal narrative suggests that simply by exposing children to the origins of their foods, they will then develop a taste for these foods – or become more willing to taste fruits and vegetables (Canaris, 1995; Davis et al., 2015; Gibbs et al., 2013; Jaenke et al., 2012; Koch, Waliczek, & Zajicek, 2006; Morris, Briggs, & Zidenberg-Cherr, 2000; Somerset, Ball, Flett, & Geissman, 2005; Somerset & Markwell, 2008). As a result, children go home and influence family shopping decisions around fruits and vegetables (Wells et al., 2018), ultimately resulting in healthier families all round. This chain of causalities does not really stand up to critical scrutiny. There are a lot of intermediary gaps between tasting something at school and changing family dietary habits.

And while we do not disagree with the value of knowing the origins of food and experiencing a range of foods, we are concerned about the way healthy eating and gardening have become equated. Children and adults alike are expected to know what 'healthy eating' means, and to eat healthily, irrespective of economic situation, cultural traditions or home food practices. But 'taste' is always associated with particular social positioning (Bourdieu, 1984); what is tasty for some is not so for others. Practitioners do not often view changing tastes via the school garden as problematic, as creating a singular, unifying 'taste' that homogenises. Perhaps as a practice situated in cultural imperialism (Hayes-Conroy, 2014). And yet, tastes of school gardens are often underpinned by a discourse that is situated in white, middle-class assumptions of health (Earl, 2020; Guthman, 2008; Hayes-Conroy, 2014; Swan & Flowers, 2015). Just like the garden advocates we saw in Chapter 1, there appears to be little awareness, amongst today's prominent school garden enthusiasts, of the histories of agriculture and enslavement. 'Putting one's hands in the soil' may not be a soothing or cathartic experience for everyone, but rather one intertwined with trauma and oppression.

Similarly, understandings of community foods, how to grow them and how to cook them, are often missing from the knowledges of enthusiastic volunteers or staff who run garden programmes. School garden programmes expose children to *particular* food knowledges – those associated with whiteness, thinness and power. While we did not explicitly encounter such issues in the school gardens we visited, we did see that healthy eating was uncritically held as an assumed knowledge.

Cairns (2018) has argued that young people become 'educational outputs of garden pedagogies' (p. 520), in the sense that they are 'cultivated' alongside the vegetables they tend. Healthy eating habits, then, are one way that children become a garden output – by changing their eating habits for the better, children become healthier, a result of their exposure to different vegetables and an increasing willingness to try new varieties. Similarly, gardens help craft children into foodies, educating them about growing their own, and knowing where their food comes from, how to produce it and then, often, how to cook it. In this way, what Cairns (2018) calls the 'magic carrot model' romanticises the transformative promise of children's experience in the garden, and obscures the need for state action and institutional change to build better, more just and more sustainable food systems (p. 519) (see also: Hayes-Conroy, 2010).

Hunger, hunger, hunger

Romanticised gardening practices often obfuscate the need for profound economic and social change. Childhood hunger and food poverty are growing and concerning phenomena in 21st-century England (Dowler & O'Connor, 2012; Garthwaite, 2017; Lambie-Mumford, 2017; Lambie-Mumford & Silvasti, 2020). Beyond healthy eating, and likes and dislikes, we encountered stories of hunger in schools. As we indicated in Chapter 3, City School was undergoing a transformation in school population demographics but poverty and hunger were still very real experiences for some of their children. Similarly, New School's children came from an economically marginalised community where poverty and hunger were real concerns – school staff were aware that the midday meal might be the only 'proper meal' (Murcott, 1982) a child could access in a 24-hour period. This was not simply a staff worry – children in focus groups also brought up the issue of hunger.

In the age of austerity, school garden projects sit within wider discourses of hunger and deprivation. Gardens are not only spaces to learn about nature or food growing, they are also important sources of food. Children at both schools were aware of food poverty. Like children's books on food banks or having to skip meals, children's familiarity with questions of hunger are illustrative of the widespread effects of current neoliberal politics, manifest in growing social inequality and lack of social mobility (Dowler & Lambie-Mumford, 2015; Loopstra, Lambie-Mumford, & Patrick, 2018).

Current literatures on the benefits of school gardens tend to focus on changing children's tastes towards new fruits and vegetables, and there is less written on the intersections of hunger, class and the garden (there are exceptions e.g. see Burke, 2005 and, of course, the literatures on feeding programmes in the Global South). However, as Williams and Brown (2012) argue, sustainability education in gardens provides children with the opportunity to make connections between seemingly distinct sustainability concepts like hunger, food and transport costs, for example. We saw these connections made by the children we spoke to (Chapters 3 and 4). We find it interesting that it was the children who raised the garden as a potential solution to hunger, and as a way to help out people less fortunate than themselves. Hunger as a general social phenomenon was not talked about by staff within the context of the garden space. The garden space was somehow separated from issues of hunger, and it was the kitchen and school dinners where the connection was made. The garden was heavily connected to deficit views of the food practices of those living in straightened circumstances. Staff tended to see hunger as an individual matter, rather than a matter of wider social and economic structures and policies. There was thus no connection with current debates about the politics of food poverty more generally and the notion of a universal right to food (Dowler & O'Connor, 2012; Lambie-Mumford, 2017; Lambie-Mumford & Silvasti, 2020).

Filling the nature deficit: building wellbeing and resilience

One of the well-known narratives surrounding school gardening programmes is the need to reconnect children to the natural world. 'Nature deficit disorder', a term coined by Louv (2010), refers to modern children's lack of interaction with the outdoor world. Louv argues that in a childhood dominated by indoor play, supervision and electronic games, children suffer from their lack of contact with the natural world, and this negatively affects their behaviour and well-being. This view is contested. In 2015, Novotny, Zimova, Mazouchová and Šorgo (2020) compared the nature experiences of children in the Czech Republic, with those from a town in Germany in 1900. They found that children had significantly more experiences of nature than children in 1900, despite the decline in farming and associated practices. Children's nature experiences tended to be within the ecosystems of gardens and forests and in time outside of school. Class, gender and race were not factors investigated, so how these impact experiences of nature is unknown. But this research raises interesting questions. How much of the concern for human–nature interaction is based in nostalgic views of the past? Is the view that children need to be 'reconnected' with nature actually a misnomer, because humans are part of nature, not separate from it (Armstrong, 2005; c.f. Mycock, 2019)? The nature deficit discourse is generally stitched together with educational narratives about the importance of exposing children to the natural world. We encountered the 'nature deficit' discourse at both our case study schools.

At both City and New School, children suggested that gardens were spaces to spend time in, without broader agendas. Children described feeling calm, being able to breathe more freely, feeling like things were going to be okay, enjoying the fresh air and being 'in' nature. For children, the garden was *more than* a space for growing vegetables or learning about the natural world. It was a space that let them become attuned to nature, and gave them the opportunity to be *with* the fresh air. City School partly understood this – they used the garden as a space for young people to unwind and relax after they had sat for standardised tests – children were given the opportunity to spend time in the garden in-between sitting exams. Whether such practices build resilience, and what kind, we do not know, but the children's opinions certainly suggest that gardens were beneficial to their well-being (see also: Malberg Dyg & Wistoft, 2018). Such unstructured encounters in the garden allowed children to build relationships with the world outside of the human-centric focus that sees humans and nature as separate entities. This is more akin to practices of Indigenous groups who take the learning principles of environmental education as the promotion and nurturing of 'wellbeing in and for the earth' – social and ecological systems are interconnected (Gaylie, 2009), and gardens provide children opportunities to experience this first hand. It is not so much a need to reconnect children to nature through the garden; rather, it is about allowing them to be in the garden.

Part of the value of gardening is that the gardener has some control over what they do. Cutter-Mackenzie (2009) and Surman and Hamilton (2019) have noted that gardens can provide agency to young people outside of formal school structures, allowing them to work to their own forms of logic and encouraging self-direction and autonomy. For Cutter-Mackenzie (2009), the purpose of gardening is not necessarily developing scientific knowledge of plants, it is the 'doing gardening' (p. 131) that is important, not the knowledge that might be gained. The very idea of unspecified school garden learning outcomes is likely to be unpopular in schools in England.

School garden experiences do not easily translate into statistics for assessment or educational audit, as we have noted – developing motor coordination via digging, becoming less scared of minibeasts like worms, spending time smelling herbs – cannot be simply measured or counted. Perhaps this is why some see gardens as less worthy of classroom-based teaching, and instruction is directed primarily to material and experiences that will 'raise individual performance and scores on standardised tests' (Moore, Wilson, Kelly-Richards, & Marston, 2015, p. 409). New School was caught directly in the dilemma of how to manage both garden learning and test results. But the freedom of younger children at both our research schools to experiment outdoors, and spend time digging in the mud, or at farm school learning about pigs and chickens, contrasts sharply with children in the upper-year groups who were revising for SATs, or improving their Mathematics and literacy scores, and whose time outside was much more strictly controlled.

The idea that the garden might provide a place to be, a space for reflection, where one can simply sit and think, or look slowly at minibeasts, or watch the wind in the leaves, or develop relationships with non-human elements (Green & Duhn, 2015) and that this might contribute to learning, is in sharp contrast to the English government's perceptions of what schooling is for. Passy (2014) has argued that in England, learning is perceived to be something that can only happen indoors, and that leaving the classroom is neither practical nor desirable in a school system focussed on 'measurable results' (p. 36). School gardens are constricted by policies that conceive of learning in particular ways, and do not provide space for experimentation, or time to explore different ways of doing things (Passy, 2014).

Access to the garden: nostalgia and adult agendas

At both New School and City School, one aspect of gardens struck us as both important and perhaps a little strange: children's access to outdoor spaces, and gardens in particular, was restricted by adults. There was much talk of the 'freedom' that being in a garden allowed children to experience and yet this 'freedom' was mitigated through adult rules and restrictions. *When* children could be in the garden, *which* children could be in the garden, *what* children could do in the garden were all restricted by adults.

At City School, the garden was not only a natural space that might be used for play. It was also conceived of as an extension of the teaching classroom. As such, children's behaviour had to be highly controlled, a practice resonant with Passy's (2014) finding that learning is something that only happens indoors, where it can be managed and monitored. Learning and behaviour outside needed to conform to the same expectations as inside. Outside though, children experienced freedom that made structured teaching more challenging, because teachers could not imagine new 'rules' for the garden. The school garden appears not to have been a space for 'safe' risk-taking, exploration and/or play, as is the philosophy underpinning Forest Schools (Knight, 2009; Mycock, 2019). The garden was a space for instruction and transmission of knowledge, even though it did not have the familiar constructs of a walled classroom. Children were aware of access restrictions, knowing that they had to have permission from an adult to visit or be in the garden.

At New School, restrictions on access were also in place. Although children could visit the farm at lunchtimes, if they so wished, they could not access the space directly – the pigs and chickens were kept behind high, locked fences. Access was only allowed in the company of the farm school teacher. Other garden spaces – the rhubarb patch, the herb garden, the wildlife pond, were also restricted to children. Play took place on the large playing field, or at the various clubs around and about the school, but the gardens remained places for more formal instruction.

Learning was not the only discourse at work in our two schools. Children could not be trusted to enter farm spaces alone and act appropriately. Restricted access to gardens thus also leads to the question of what are seen as suitable activities for children of particular ages – perhaps children were not mature enough to take responsibility for themselves (the 'stages' discourse of universal, age-related linear development as per Spencer, Montessori and, later, Piaget). Perhaps gardens were seen as risky places, and access without a teacher constituted a violation of health and safety regulations. Garden-as-learning could easily and neatly stitch together with other dominant schooling discourses.

We found the restriction of access interesting. So much of the talk from adults we spoke to was focussed on children experiencing natural spaces, being able to explore and encounter nature, and yet the reality of the school garden was one of restriction and control. We are not arguing here that discipline, the practice of creating and keeping social order, is unimportant. It clearly is. Children cannot learn without rules and consequences. We are however noting the differences between garden discourse and that shaping other forms of outdoor activity, such as outdoor education, where controlled risk-taking is encouraged. We also note cultural differences between the English norms of youthful orderliness and the discourse of exploration and self-management underpinning Nordic outside-learning traditions (e.g. Bentsen, Mygind & Randrup, 2009; Dahlberg, Moss & Pence, 2006; Waite, Bølling & Bentsen, 2016).

Our two schools' concerns about control and access are also related to the ways in which they understood resilience. Resilience, often referred to as the capacity to 'bounce back' from challenges, is a state, condition and practice (Knight, 2007) which is 'nurtured' in contexts – family, community, school. Researchers (e.g. Johnson, 2008; Oswald, Johnson and Howard, 2003) point to the important contextual contributions made to children's resilience by the school; teachers can do a lot to help children develop their sense of self, self-belief and self-efficacy, to take risks and learn from them, to take responsibility for themselves and others, and to seek support for themselves when needed, and give support to others. However, pastoral concerns are often separated discursively from questions of school discipline. The potential tension between resilience and the imposition of strong external rules is obscured. The question of whether children will develop independence and self-managing practices if they are highly restricted is thus not on the agenda. This was particularly the case we noted at City School.

Control and access also sat alongside nature deficit concerns already mentioned. Louv's (2010) concerns for children's lack of engagement with the natural world harks back to times when children rode bicycles all afternoon in the streets, or explored the local woods, or climbed trees; in short, when children experienced much more freedom to roam and explore their local communities than is commonplace today. Jorgenson (2013) and Wake (2008) have both noted the tendency of school gardens to be places where nostalgia and romanticised notions of nature come to the fore. Indeed, Wake (2008) argues, children's gardens reflect and perpetuate adult agendas. Heavy nostalgia in school gardens is perhaps not so surprising when considered alongside much of the discourse that defines the 'alternative' food movement and motivates activists set on 'doing good' in communities. Knowing where food comes from is something people 'just knew' in times gone by, particularly during the Second World War's Dig for Victory campaigns taken up in the UK and US (Potter & Westall, 2013). As we discussed in Chapter 1, an early focus for school gardens was to provide fresh air and exercise for city children, and it seems less has changed around these priorities than we might think. Our school gardens were places of learning, first and foremost, that might become spaces of freedom at a later stage. Gardens were viewed as fundamentally important to *improving* children, in controlled conditions, and eventually (through changing their preferences for fruits and vegetables, say) their communities (Robin, 2001).

Problematising school gardens

School gardens can be wondrous spaces, a literal jungle of plants, minibeasts, insects, flowers and space. But as we have discussed through this chapter, school gardens are more often framed as solutions to various social problems including childhood obesity, increasingly sedentary lifestyles, lack of nature connections and stress. The way that problems are framed then narrows the potential solutions to those problems, and shapes the ways children can be and become (Ball, 1993; Hacking, 2006). In this section, we discuss how the two different schools in our research took up various discursive elements to shape their garden narratives. Through this description, we highlight the assemblages schools use to conceive of the 'problem' for which a school garden is the, or a partial, solution. Gardens become materialisations of bigger school decisions and perceived purposes.

New School largely drew on discourses related to well-being and resilience in gardens, framed as part of an agenda on sustainable development within a wider narrative of ecological crisis. The focus was whole school change which began with the development of more sustainable buildings, and then evolved to include the outdoor spaces of the school. The focus on sustainable development included connections to seasonal changes (through harvest festivals, changing indoor displays), learning about raising animals, spending time outdoors and growing vegetables.

For New School, the 'problem' that might be addressed by school gardens was the question of potential ecological collapse, and the way young people might be connected to sustainable living practices through garden practices. Acutely aware of the ecological crisis, and ahead of his time, the head teacher at New School was pioneering in his approach to the school environment and the curriculum. Rather than conceiving of the garden as a separate part of the school, the garden at New School was integrated into the school community and, to a certain extent, into the curriculum. Until the change of head teacher in 2012, the garden spaces were used as part of the school day – teachers tended the garden plots with their classes, took children outside when it snowed (snow was such a rare occurrence in the school's location) and children of various ages participated in the farm school.

Outside and indoor spaces were connected and fluid. Not only did the new building have floor-to-ceiling windows that overlooked the green fields, drawing the outside in, but time spent outside was often lengthy, sometimes most of a day could be spent learning outside. The garden was initially spoken and animated through a discursive assemblage which foregrounded well-being, resilience and sustainability. Later, as the farm school became the focus of outdoor activity, and as time in the garden was cut shorter due to curriculum pressure, the assemblage shifted to incorporate food sustainability, and knowing where food comes from, in addition to the more general focus on sustainable development. Well-being and resilience became less part of the garden assemblage as the school changed. The garden and farm were increasingly seen as places to teach children about food, rather than spaces to spend time for the pleasure of being/becoming.

City School framed its garden as a health promotion tool, drawing on health and obesity concerns. Discourses of healthy eating, foodieness and obesity combined with resilience served to frame the problem of children's health at City School and were integral to its deficit view of the local community. The staff's welfarist view of a community in need, one where children needed to become something other than what they were, neatly stitched together with the foodie discourse familiar to new middle-class incomers. Perhaps foodieness even acted to make the school 'feel' more like one they would want their children to attend. This new assemblage was effected through the significant involvement of external stakeholders in the conception and initial development of the project – stakeholders who were in the restaurant industry, and whose own agendas were centred on promoting food knowledge and foodieness within the school.

City School is an example of a power struggle exerted over and through a social and educational deficit-foodie discourse assemblage. Different stakeholders initially cooperated to shape the purpose and practices of the garden. External stakeholders were committed to teaching children about food. The school management team agreed but also wanted to create a space to enhance the well-being of their children, provide them with outdoor access, improve their health and improve the school grounds. City School's foodieness is best represented by the food festival the school hosted to raise funds for the garden. This festival connected a financially strapped community with top restaurants and chefs, many with their own cookbooks and/or television programmes. School management welcomed the funds raised, and they too wanted to engage the local community and create a space that was open to different experiences. However, they faced a challenge in developing stronger curriculum connections and a more inclusive narrative. The garden, at least while we were researching it, remained a separate entity from the rest of the school. Permissions to go to the garden were restricted, especially for children. Foodie discourse was largely sutured to the educational manifestation of deficit discourse via the kitchen and food education.

In contrast to New School, City School garden had yet to find a place within the wider school ethos. The discursive assemblage shaping its initial creation, and framing integration into the school, had yet to be made central to its wider learning goals and programme. There were considerable tensions just below the surface. How did the food festival sit within inclusive school values, when it focussed on high-end restaurants and chefs, most of whom were white and middle-class? In a school in a mixed neighbourhood, with children from diverse backgrounds, did such a high-profile event serve to foodwash diverse cultural practices, traditions and cultures of food? What impact would the food festival have on how children perceived the garden space? How might it impact on the community's perceptions of the school – were parents to be fundraisers, volunteers or possess useful funds of gardening knowledge?

We see City School as an example of a garden struggling to settle its discursive and material place. Was the garden to teach deprived young people about (white, classed) food practices? This was certainly the way external stakeholders in the garden thought about the problems the garden might address (although they may have been unaware that this is how they were framing the problem). Or was it to be a space where children could build resilience, improve their well-being and spend time outdoors? How important were the vegetables grown in the garden to its wider purposes? Management was strongly focussed on trying to integrate the garden into the wider school ethos but this was perhaps made more difficult when the garden was simultaneously being moulded by an alternative problem and solution. How could 'taste' and being resilient come together?

The silences in garden talk

As we have seen throughout this chapter, school gardens are often framed by adult concerns regarding health and well-being, access to the natural world, and consumption of fruits and vegetables. Children's choices and use of outdoor spaces, Wake (2008) argues, do not always conform to adult expectations and ideas. These struggles are very visible in the conversations we had with teachers – where we witnessed the desire for children to behave and be seen to behave in order to learn, in a space also designed to encourage their freedom and exploration.

In this section, we explore what is left out of current conversations of children's gardening. Who does not garden? What gets excluded in a school garden? As we have shown throughout this chapter, school gardens and gardening are framed through particular discourses of understanding and justification: gardens teach young people about health by encouraging them to eat or try new vegetables; gardens provide young people with spaces to play (encouraging physical activity); gardens encourage children to reconnect with the natural world. The school garden is seen as a potential solution to all these problems. And yet the way these problems are framed, also serves to frame the ways gardens are grown, used and embedded within a school community. We focus on three key silences: obesity, cultural assumptions about vegetables and classed assumptions about children's experiences.

Obesity

The 'problem' of 'obesity' which the garden addresses through teaching about healthy eating, is disputed amongst academics and health professionals, as we have already noted. There is no consensus as to whether there is an obesity 'epidemic' nor as to what are the best ways to reduce body weight (or whether this is even advisable) (Earl, 2018; Gard & Wright, 2005; Lupton, 2013). The garden is situated within obesity debates, even though these can remain largely hidden. Solutions to the obesity crisis have been criticised for being couched in middle-class, white ideas of health, thinness and body size desirability (Guthman, 2011), and when school garden advocates talk about encouraging children to try new vegetables, they are situating the garden with the desire for thinness. There is also the question of causality argument: children will try more vegetables and then be receptive to trying more, this will lead them to develop a liking for vegetables. In turn, they will influence their families' food choices by encouraging more vegetable consumption. Families will therefore be healthier and thinner. What is often not added is that responsible self-managing families will not be a burden on underfunded public health services.

Cultural assumptions about vegetables

The types of vegetables grown in school gardens are not often specified by gardeners – they are referred to generically as fruits and vegetables. We were interested that most of the vegetables we saw in school gardens tended to be those belonging to traditional, white, British foodways – tomatoes, carrots, peas, potatoes, strawberries. School gardeners did not talk about growing vegetables familiar to the children within the school – children who came from different cultural and immigrant backgrounds for whom perhaps carrots and peas may be 'foreign' foods. Do school gardens grow okra? Aubergines? Sweet potatoes? White radishes? Pak choi? What do school gardens say to children about the myriad different vegetables grown around the world that are then used in different cuisines? Our time in gardens suggests that British fruits and vegetables are the focus, leaving out all the other fruits and vegetables cultures eat, building Anglo 'taste' and implicitly questioning the acceptability of eating such foods.

Classed assumptions of children's experiences

The teachers we spoke to assumed that children from underprivileged backgrounds did not have the opportunity to spend time outdoors, amongst the wilds of nature. This was often framed as children playing video games indoors, when not in school – an implicit critique of parents. The locked-inside assumption obscures that fact that some neighbourhoods are not safe for small children (busy roads, lack of playgrounds), and keeping children indoors may, in fact, be an act of caring parenting. Additionally, while staying inside may have been true for some children in our two schools, it was certainly not true for all the children we spoke to, many of whom visited grandparents with gardens or played in the local parks. But the assumptions underlying this teacher-belief were built on assumptions of deficit.

Spending time outside not only had health effects. It was also to do with becoming a good citizen. Gardens were seen as vehicles for turning children into responsible consumers – future citizens who would protect the natural world, care for it and, quite possibly, save it. The assumption here was that children from homes where money was short, housing was cramped and public health outcomes were poor, needed more help to become orderly citizens. They did not have access to the 'normal' experiences at home, so the school had to compensate for this lack.

Cairns (2018) argues that 'by appealing to a universal child as output in a placeless garden, the rhetoric of effects obscures the uneven landscapes in which children's subjectivities are formed' (p. 521). Victorians sought to 'cultivate' the wild, unruly child, and there is a similar narrative underpinning current school gardens related to civilising children – whether that civilising refers to their food choices and behaviours, their self-control in garden spaces, or their ability to control (and shape) their bodies. Such outcomes are tied to conceptions of children as innocent and pure, incapable of moulding their own environments and ignoring the inherent injustices, histories and legacies that shape their encounters with their surrounding environments (Cairns, 2018). Labouring in the garden is seen as a panacea for a wide array of social deficits, just as occurred in the early 20th century during the first school garden movement.

In Chapter 6, we will focus on growing a school garden as whole school change, within the context of education/public policy, curriculum requirements and wider social contexts.

References

Armstrong, J. C. (2005) En'okwin: Decision-making as if sustainability mattered. In M. Stone & Z. Barlow (Eds.), *Ecological literacy: Educating our children for a sustainable world.* San Francisco, CA: Sierra Club Books, 11–17.

Ball, S. J. (1993.) What is policy? Texts, trajectories and toolboxes. *Discourse: Studies in the Cultural Politics of Education,* 13(2), 19–17.

Beery, M., Adatia, R., Segantin, O., & Skaer, C. F. (2014). School food gardens: Fertile ground for education. *Health Education,* 114(4), 281–292.

Bentsen, P., Mygind, E., & Randrup, T. B. (2009). Towards an understanding of *udeskole*: Education outside the classroom in a Danish context. *Education 3–13, 37*(1), 29–44.

Bourdieu, P. (1984). *Distinction: A social critique of the judgement of taste.* (R. Nice, Trans.) Boston, MA: Harvard University Press.

Burke, C. (2005). Contested desires: The edible landscape of school. *Paedagogica Historica, 41*(4&5), 571–587.

Cairns, K. (2018). Beyond magic carrots: Garden pedagogies and the rhetoric of effects. *Harvard Education Review, 88*(4), 516–538.

Canaris, I. (1995). Growing foods for growing minds: Integrating gardening and nutrition education into the total curriculum. *Children, Youth and Environments, 12*(2), 135–142.

Cutter-Mackenzie, A. (2009). Multicultural school gardens: Creating engaging garden spaces in learning about language, culture, and environment. *Canadian Journal of Environmental Education, 14*, 122–135.

D'Abundo, M. L., & Carden, A. M. (2008). Growing wellness: The possibility of promoting collective wellness through community garden education programmes. *Community Development, 39*(4), 83–94.

Dahlberg, G., Moss, P., & Pence, A. (2006). *Beyond quality in early childhood education and care: postmodern perspectives* (2nd ed.). London: RoutledgeFalmer.

Davis, J. N., Spaniol, M. R., & Somerset, S. (2015) Sustenance and sustainability: Maximising the impact of school gardens on health outcomes. *Public Health Nutrition, 18*(13), 2358–2367.

Dimbleby, H., & Vincent, J. (2013). *The school food plan.* http://www.schoolfoodplan.com/wp-content/uploads/2013/07/School_Food_Plan_2013.pdf: London.

Dowler, E., & Lambie-Mumford, H. (2015). How can households eat in austerity? Challenges for social policy in the UK. *Social Policy and Society, 14*(3), 417–428.

Dowler, E., & O'Connor, D. (2012). Rights-based approaches to addressing food poverty and food insecurity in Ireland and the UK. *Social Science and Medicine, 74*(1), 44–51.

Earl, L. (2018) *Schools and food education in the 21st century.* Abingdon: Routledge.

Earl, L. (2020). 'Who likes fish? And I don't mean fish fingers!' Taste education, culinary capital, and distinction in a rural middle-class primary school. *Cambridge Journal of Education, 50*(1), 77–93.

Gard, M., & Wright, J. (2005). *The obesity epidemic: Science, morality, ideology.* Abingdon: Routledge.

Garthwaite, K. (2017). *Hunger pains: Life inside foodbank Britain.* Bristol: Policy Press.

Gaylie, V. (2009). *The learning garden: Ecology, teaching and transformation.* New York: Peter Lang.

Gibbs, L., Staiger, P., Johnson, B., Block, K., Macfarlane, S., et al. (2013). Expanding children's food experiences: The impact of a school-based kitchen garden program. *Journal of Nutrition Education and Behaviour, 45*(2), 137–146.

Green, M., & Duhn. I. (2015). The force of gardening: Investigating children's learning in a food garden. *Australian Journal of Environmental Education, 31*(1), 60–73.

Guthman, J. (2008). Bringing good food to others: Investigating the subjects of alternative food practice. *Cultural Geographies, 15*, 431–447.

Guthman, J. (2011). *Weighing in: Obesity, food justice and the limits of capitalism.* London: University of California Press.

Hacking, I. (2006). Making up people. *The London Review of Books*, August, 1–12.

Hayes-Conroy, J. (2010). School gardens and 'actually existing' neoliberalism. *Humboldt Journal of Social Relations, 33*(1), 64–96.

Hayes-Conroy, J. (2014). *Savoring alternative food: School gardens, healthy eating and visceral difference.* London: Routledge.

Jaenke, R. L., Collins, C., Morgan, P., Lubans, D. R., Saunders, K. L. et al. (2012). The impact of a school garden and cooking programme on boys' and girls' fruit and vegetable preferences, taste rating and intake. *Health Education and Behaviour, 39*(2), 131–141.

Johnson, B. (2008). Teacher-student relationships which promote resilience at school: A micro-level analysis of students' views. *British Journal of Guidance & Counselling, 36*(4), 385–398.

Jorgenson, S. (2013). The logic of school gardens: A phenomenological study of teacher rationales. *Australian Journal of Environmental Education, 29*(2), 121–135.

Knight, C. (2007). A resilience framework: Perspectives for educators. *Health Education, 107*(6), 543–555.

Knight, S. (2009). *Forest schools and outdoor learning in the early years.* London: Sage.

Koch, S., Waliczek, T. M., & Zajicek, J. M. (2006). The effect of a summer garden program on the nutritional knowledge, attitudes, and behaviours of children. *HortTechnology, 16*(4), 620–625.

Lambie-Mumford, H. (2017). *Hungry Britain: The rise of food charity.* Bristol: Policy Press.

Lambie-Mumford, H., & Silvasti, T. (Eds.) (2020). *The rise of food charity in Europe.* Bristol: Policy Press.

Lineberger, S. E., & Zajicek, J. M. (2000). School gardens: Can a hands-on teaching tool affect students' attitudes and behaviours regarding fruits and vegetables. *HortTechnology, 10*(3), 593–597.

Loopstra, R., Lambie-Mumford, H., & Patrick, R. (2018). *Family hunger in times of austerity: Families using food banks across Britain.* SPERI British Political Economy Brief No. 32. Sheffield: Sheffield Political Economy Research Institute.

Louv, R. (2010) *Last child in the woods.* New York: Atlantic Books.

Lupton, D. (2013). *Fat*. London: Routledge.

Malberg Dyg, P., & Wistoft, K. (2018). Wellbeing in school gardens – the case of the Gardens for Bellies food and environmental education program. *Environmental Education Research, 24*(8), 1177–1191.

Moore, S. A., Wilson, J., Kelly-Richards, S., & Marston, S. A. (2015). School gardens as sites for forging progressive socioecological futures. *Annals of the Association of American Geographers, 105*(2), 407–415.

Morris, J., Briggs, J. M., & Zidenberg-Cherr, S. (2000). School-based gardens can teach kids healthier eating habits. *California Agriculture, 54*(5), 40–46.

Murcott, A. (1982). On the social significance of the 'cooked dinner' in South Wales. *Social Science Information, 21*(4/5), 677–696.

Mycock, K. (2019). Forest schools: Moving towards an alternative pedagogical response to the Anthropocene? *Discourse: Studies in the Cultural Politics of Education, 41*(3), 427–440.

Novotny, P., Zimova, E., Mazouchová, A., & Šorgo, A. (2020) Are children actually losing contact with nature, or is it that their experiences differ from those of 120 years ago? *Environment and Behavior*, 1–22.

Oswald, M., Johnson, B., & Howard, S. (2003). Quantifying and evaluating resilience-promoting factors: Teachers' beliefs and perceived roles. *Research in Education, 70*(1), 50–64.

Passy, R. (2014). School gardens: Teaching and learning outside the front door. *Education 3–13 International Journal of Primary, Elementary and Early Years Education, 42*(1), 23–38.

Potter, L., & Westall, C. (2013) Neoliberal Britain's austerity foodscape: Home economics, veg patch capitalism and culinary temporality. *new formations: a journal of culture/theory/politics, 80*(80), 155–178.

Robin, L. (2001). School gardens and beyond: Progressive conservation, moral imperatives and the local landscape. *Studies in the History of Gardens and Designed Landscapes, 21*(2), 87–92.

Somerset, S., Ball, R., Flett, M., & Geissman, R. (2005). School-based community gardens: Re-establishing healthy relationships with food. *Journal of the Home Economics Institute of Australia, 12*(2), 25–33.

Somerset, S., & Markwell, K. (2008). Impact of a school-based food garden on attitudes and identification skills regarding vegetables and fruit: A 12-month intervention trial. *Public Health Nutrition, 12*(2), 214–221.

Surman, E., & Hamilton, L. (2019). Growing consumers through production and play: A phenomenological exploration of food growing in the school foodscape. *Sociology, 53*(3), 468–485.

Swan, E.,& Flowers, R. (2015). Clearing up the table: Food pedagogies and environmental education – contributions, challenges and future agendas. *Australian Journal of Environmental Education, 31*(1), 146–164.

Waite, S., Bølling, M., & Bentsen, P. (2016). Comparing apples and pears?: A conceptual framework for understanding forms of outdoor learning through comparison of English Forest Schools and Danish *udeskole*. *Environmental Education Research, 22*(6), 868–892.

Wake, S. J. (2008). In the best interests of the child: Juggling the geography of children's gardens (between adult agendas and children's needs). *Children's Geographies, 6*(4), 423–435.

Wells, N. M., Meyers, B., Todd, L., Henderson, C. R., Barale, K., et al. (2018). The carry-over effects of school gardens on fruit and vegetable availability at home: A randomised control trial with low-income elementary schools. *Preventive Medicine, 112*, 152–159.

Williams, D. R., & Brown, J. D. (2012) *Learning gardens and sustainability education.* London: Routledge.

6 Why garden in schools?

We titled our book *Why garden in schools* because we wanted to show that the reasons for gardening in school varied over time and place. We also wanted to show that the reasons for establishing school gardens led to different kinds of practices, benefits and challenges. In tracing the history of school gardens, we saw not only a range of answers to the 'why garden?' question but also a 'boom and bust' history which deserves further attention.

Early school gardens were established because it was believed that gardening would instil the value of hard work and build moral character. In the late 19th and 20th centuries, school gardens were additionally seen as learning about agriculture and Horticulture, a valuable life skill and/or a post-school vocation. Gardens were also tied to various nation-state concerns – representing the imaginary of a nation through its plants, growing food during a global crisis, providing a redemptive practical skill for young people who might otherwise direct their energies elsewhere, promoting healthy domestic exercise and even introducing new crops.

Our two portraits represent the two most common current answers to the question 'why garden?' The first is the kitchen garden designed to address health issues and promote a particular foodie lifestyle. The second is the eco-school which attends to sustainable development. Each of these garden programmes draws on a wider social problematisation – population health problems and lack of understanding about food, and the ecological crisis. Each was, we suggested, a particular kind of discursive assemblage. Assemblages take material and relational forms – in our two portraits, we saw the same set of garden discourses in play but with different emphases and framings. This final chapter considers some further key aspects of the two school gardens we identified.

We focus in this concluding chapter on more pragmatic questions about sustaining school gardens. We examine three areas: (1) the school garden and the curriculum, (2) the practical challenges of school gardening and (3) the garden as a school change initiative. These areas emerged as important in both our literature and empirical work. We thus draw in this chapter on our historical and literature analysis, our two garden portraits and the literatures on school change (e.g. Ainscow, 2015; Datnow, Hubbard, & Mehan, 2002; Day, Gu, & Sammons, 2016; Earley & Greany, 2017; Elmore, 2004; Fullan, 2011; Hargreaves, 1996; Hargreaves & Goodson, 2006; Lingard, Christie, Hayes, & Mills, 2003; Macbeath & Dempster, 2009; Thomson, 2011; Wrigley, Thomson, & Lingard, 2011), as well as our own related research (e.g. Hall & Thomson, 2017; Thomson, Jones, & Hall, 2009) and professional experiences.

The school garden and the curriculum

If the school garden is the answer to a social or educational problem, the particular problematisation does more than create the warrant for the garden. It also establishes a link to the educational programme of the school, and to pedagogy and curriculum. There is then a question of whether the garden can move beyond its initial curriculum linkage to become a site for cross-curriculum teaching and learning.

Stitching the garden into the curriculum

We showed in Chapter 1 that Comenius, an educator in a religious order, saw the garden as a place for religious and moral instruction, understood at the time as a discrete area of knowledge and instruction. The view that the garden was a curriculum domain in its own right persisted. In the newly compulsory schools of the late 19th and early 20th century, gardens had a particular place. Schools were expected to teach children and young people the wherewithal to live law-abiding lives and to undertake productive work – children were seen as earners, already often contributing to family incomes, not as learners (Mayall & Morrow, 2011). Gardening was thus often a discrete school subject in the elementary years offering practical horticultural knowledge as well as social education; in secondary schools, Gardening was a vocational subject.

In the US, where school gardening was systemically supported, garden-as-discrete-subject development work was largely undertaken outside schools – nationally, and at state and district level. Pioneering garden consultants and advisers wrote manuals for schools; schools also relied on the advisory support and pressure from their school districts to successfully implement garden programmes. In the UK, however, school gardens seemed more a matter for individual schools and perhaps newly established local authorities (we say more on this later). External systemic support for gardens was particularly strong in both countries during the two world wars and the Depression. But when social problems and problematisations changed, policy priorities shifted and external supports fell away, as did the school gardens and the garden-as-discrete-subject.

By contrast, and over the same time period in the 19th–20th centuries, the subject 'Nature Study' became an integral part of the primary science curriculum. Nature Study was systemically supported – there were

- strong links between university science researchers and Nature Study teachers forged through the Nature Study movement;
- primary teachers trained in Nature Study as part of their initial preparation;
- mandated textbooks dedicated to Nature Study, building cumulative biological and botanical knowledges over each grade; and
- early school inspectors who expected to see a Nature Study table or ledge in every classroom – Nature Study was as taken for granted in primary education as a handwriting lesson (Bittner, 1919).

Sometimes school gardens were integral to Nature Study but often they were kept separate; while all teachers were responsible for Nature Study and Science learning, most often one teacher or staff member took responsibility for the school garden, even if all classes participated. Our reading of curriculum histories suggests that, in the school knowledge hierarchy, the subject Science was more important than gardening. Nature Study thus persisted when gardens did not.

Today's school gardens also have a fragile curriculum connection. It is not clear whether they are a subject in their own right or part of another subject. And if the latter, which one?

The kitchen garden – the type that City School aspired to and the Edible School Yard and Stephanie Alexander Kitchen Garden programmes promote and support – is primarily connected to health education via the provision of healthy school meals (US and UK), healthy school canteens (Australia) and formal health education lessons. Health Education itself is part of a larger 'subject', called Personal,

Social and Health Education (PSHE) in the UK. PSHE includes Physical Education as well as Health Education. The Health Education part of PSHE also has to include Sex and Relationships Education and general social and emotional development as well as lifestyle choice. While compulsory until the end of junior high school, PSHE is one of the subject areas that fights for time and parity of esteem with the 'core' subjects such as English and Maths. PSHE teachers in secondary schools are not always trained in the subject and may not feel at all at home with the idea of gardening.

The eco-school garden – the type that New School successfully maintained – is not necessarily linked to any particular subject area. In primary schools, Education for Sustainable Development (ESD) is a cross-curriculum area which works with concepts from both Science and Social Science. It is likely to be part of a thematic pedagogical approach which, in eco-schools, is supported by external advisers and materials, and generally an internal middle leader and/or specialist staff. In secondary schools, ESD may be part of the subject Geography with elements taken up in the sciences, or vice versa. However, a school garden is generally only one of a series of strategies included in ESD. As the example of New School shows, ESD is also often literally built into the school architecture, and into its energy, water and waste systems, as well as into the curriculum.

The garden as a location for cross-curriculum learning

There are ready connections to be made between school gardens and all subjects in the curriculum. From Comenius on, early garden pioneers saw such links, as did the later European and US school garden movements.

But making connections across the curriculum, as it is most often organised, is not easy. Development and planning for any cross-curriculum area means establishing points of complementarity across knowledge domains. In the case of gardening it also means:

- deciding which disciplinary concepts are amenable to animation through gardening;
- determining sequencing and pacing of key garden ideas over time and seasons;
- designing practical activities which support conceptual progression; and
- formulating appropriate assessment tasks which are amenable to mapping learning back into disciplines and subject areas.

The first garden advocates did this curriculum development work themselves. But in the highly centralised school systems which remained the norm for most of the twentieth century, such curriculum development work was often undertaken outside of schools. In Chapter 1, we saw that systemic support for the US school garden programmes led to curriculum frameworks and materials generated at a national level as well as by 'experts' at state and/or district level. However, today's school systems operate very different curriculum regimes which play out in school garden curriculum development.

Australia, for instance, has a national curriculum organised into curriculum areas. States adapt this national curriculum, often adding in cross-curriculum themes, such as ESD, and sometimes more detailed outcomes and levels of attainment. But while national and state curricula offer an explicit and common frame, cross-curriculum integration in particular is often precarious (Salter & Maxwell, 2016). Research shows there can be a considerable gap between the ideals espoused in a national curriculum, or programme, and what schools actually do. Investigations into the introduction of new curricula framings suggests that there is often a problem at the national level – a lack of coherence and mixed messages about knowledge (Priestley & Sinnema, 2014) and/or a failure to communicate the pedagogical theory underpinning change (Taylor, 2016). Schools are thus not always able to translate and enact new curriculum frames.

The potential gap between the curriculum ideal and what Australian schools might do has been addressed by some local garden programmes. For example, the Stephanie Alexander Kitchen Garden programme has produced a range of curriculum materials for schools. The materials link garden learning to specified national curriculum outcomes across a range of subjects and year levels. Schools find such materials useful – they show how cross-curriculum learning can be organised and they encourage teachers to take first steps in that direction. But as we saw in Chapter 2, prepared materials also have potential downsides:

- schools may decide to simply adopt externally developed materials rather than adapt them for their particular students
- materials may exclude local and community gardening experiences, particularly Indigenous and/or intercultural knowledges and practices
- teachers may actually have more sophisticated pedagogical and curriculum repertoires than are embedded in external materials – in this case, materials could be seen as part of a more general deskilling agenda.

England, where are our two portrait schools are based, has a different kind of curriculum regime. There is a national curriculum with specified outcomes and levels. However, the main driver of what schools do comes from the inspection regime and narrow pedagogical foci. Government policy, until very recently, focussed on the lesson, rather than overall curriculum development, and it still promotes subject-based teaching for whole classes. Schools are also subject to skills testing and the results are used to trigger inspections which can result in drastic intervention (forced academisation or closure) (Ball, 2018).

Even when schools in England want to take a cross-curricula approach, there are still sometimes problems. We saw in Chapter 3 that some City School teachers struggled with the garden in the initial stages, seeing the need to improve literacy and numeracy as a more important priority. New School initially saw the garden as integral to making ESD cross-curriculum, and was more successful in integrating it across the school. However, test results and inspection placed intense pressure on the time spent on gardening and its place in the curriculum. City School and New School are not alone in their struggles to put the garden in its (curriculum) place.

The international school garden evaluations we reported in Chapter 2 noted that embedding the garden in mainstream curriculum was often difficult for schools. This was in part because there are significant challenges for teachers in moving away from a familiar approach to another, as well as community scepticism. In addition, there are also questions of reputation and risk involved in making the school garden a locus for cross-curriculum learning. This is particularly the case in England where a divisive debate has positioned cross-curriculum approaches as skills and process focussed, with shallow intellectual content (Boyle & Bragg, 2008; Thomson & Riddle, 2019). We do see some basis for these concerns; there *is* documentation of project-based learning, particularly in primary schools, which lacks rigour and purpose (Thomson, Hall, & Jones, 2012). But there are, of course, many examples where negotiated individual and group work is far more effective than a steady diet of teacher-directed, subject-based instruction (Galton, Hargreaves, & Pell, 2009). Nevertheless, it is a confident school sure of its test results that takes on gardening, or indeed, any cross-curriculum initiative.

As we wrote this book, we began to wonder if school gardens are the most long-standing example of the difficulties of inserting cross-curriculum courses and activities into the even longer-standing subject-based organisation of schools. As we saw in US garden histories, learning woven across a linear timetable can easily be seen as taking time away from what is most important, what is examined and what teachers were trained to do and identify with. And even when they are seen as important, cross-curriculum or project-based learning often begins with much enthusiasm, but then dies away when early advocates move on, or when implementation difficulties become too great (see Hayes, Mills, Christie, & Lingard, 2005; Mills & McGregor, 2016 for a contemporary story of an ambitious project to deconstruct subjects).

But teacher knowledges and subjects are not the only potential stumbling blocks for well-intended school garden plans. Time is also an issue. While an individual teacher in a primary school *can* organise a regular time for an integrated subject and ensure it complements and supplements teaching in core subjects, they still need time for planning. Time is hard to come by in schools. It is organisationally easier for schools to run a short-term project or theme than it is to sustain a permanent place for an activity which has multiple connections across subjects and age groups and which requires considerable expertise, time and money. How much easier to have a garden subject and a specialist garden teacher!

But of course, curriculum is more than a matter of subjects. It is also enacted, it is a practice. Animating a curriculum is complex and more than simply a planning matter. Questions of discipline and order, beliefs about the development of children and purposes of schooling more generally, all affect how any activity, including gardens, are made pedagogical. As we saw in both New and City Schools, the degree to which a garden might be a site for learning was profoundly shaped by discourses of risk, social order and age-appropriate agency. School gardens often sit at the intersection of a variety of discourses and practices, and these too can prevent or enable any curriculum goals attached to the garden to be realised or stymied.

We wonder how much the stubborn 'grammar of schooling' (Tyack & Cuban, 1995; Tyack & Tobin, 1994) – long-term beliefs and organisational structures – work to eventually squeeze gardening out of everyday classroom routines. Do the combined difficulties of finding a subject to cling to and making secure connections across the curriculum, help to explain the in-and-out-of-favour history of school gardens? Is outside learning just too difficult to control and assess? Is it too much to ask a teacher to add into their already stretched everyday work? How much were, and are, the multiple purposes of and linkages for school gardens a weakness, rather than a strength?, we ask ourselves.

Challenges of school gardening

Previous chapters reported some of the practical issues that schools faced when setting up and maintaining school gardens – funding, space, seasons, maintenance, enthusiasm and voluntary labour. None of these factors is insurmountable but each can contribute to making a school garden difficult for those schools that have one, and act as a disincentive for schools that would like to establish a garden.

Space

Schools that have no surplus land may find simply getting enough space to garden an issue. This is not a new problem. In Chapter 2, we mentioned Susan Sipe's (1909) account of an East End London school which used a graveyard for its garden as there was no room in the meagre grounds. Today UK schools often have access to allotments and community gardens. But using non-school-based land requires schools to make a lot of risk management assessments and supervision arrangements. Many schools do garden in exactly this way, but some may find the combination of time/distance to walk to the allotment and risk management too burdensome a task.

Schools that *do* have enough land have to examine the initial logistics of setting up a garden – these may range from landscape design and moving play equipment to digging up asphalt, organising weed eradication and preparing the ground for planting. This preparation may involve heavy machinery (which has funding implications) and/or working bees of volunteer parents. And organising all of the preparatory work may just be too much work for busy schools under pressure to take on.

Funding

In the UK, and elsewhere, funding is one of the biggest challenges for the school garden. School gardens require initial infrastructure, be that a greenhouse, equipment and/or raised beds. They also have ongoing costs – gardens need compost, particularly if local soils prove contaminated or unsuitable, and seeds and/or young plants, which must either be bought or procured via donations from enthusiast parent-gardeners or garden centres. Joining with an external gardening organisation can also mean a considerable investment – apart from the initial joining fee, there are ongoing expenses including access to training and additional resources. As we saw in Chapter 3, schools with well-connected parents can sometimes overcome funding constraints by running festivals and raising money but, as we indicated, this fundraising still has costs in terms of staff time and volunteering. Schools that do not have this luxury need to develop longer-term budgeting plans in order to sustain their gardening programme.

Seasons

The seasons often prove to be the greatest stumbling block in maintaining a school garden. When vegetables and fruit reach their peak ripeness, flavour and colours, schools are generally on holiday – children are absent, unable to participate in and celebrate the harvest. This is demotivating for students. Why garden if you cannot harvest the crop?

And schools do not necessarily know which plants crop early and late or have access to early/late cropping seeds or seedlings. The seasonal challenge can be countered somewhat by growing vegetables all year round, having a greenhouse, composting heavily, using shadecloth and automatic watering systems, growing fruit trees which crop later in the season, calling on a roster of volunteers, and harvesting in September (UK) / February (Australia) just as schools return. However, the harvest season is always going to be a challenge. Indeed, schools that serve agricultural communities may find that students from farming families are absent from schools just at the time when their knowledges and experience would be most useful.

Maintenance

Gardens require consistent upkeep. As we saw in Chapter 1, it was not uncommon for US children to spend one or two hours a week in the garden weeding, pruning and trimming. Some garden advocates saw weeding as a job for the girls while the boys did the heavy digging. The physical labour involved in tending to the garden was seen as instilling the moral value of hard work. But hard labour and taking time away from lessons is likely to be unacceptable in today's schools. Schools therefore often use strategies such as lunchtime and after-school garden clubs and working bees, and get much of the heavy work done by a grounds or garden staff member equipped with appropriate machinery and tools.

Enthusiasm

Maintaining teacher/pupil enthusiasm for gardening is another challenge. As we mentioned earlier in this chapter, the national policy focus is on in-classroom learning leading to specific results and test scores. Teachers must therefore make decisions around when to visit the garden, how often, for how long and whether it is possible or reasonable to sacrifice time spent on (measurable) learning. Too little engagement will not sustain children's enthusiasm and this, in turn, makes it difficult for teachers to maintain a garden programme.

Over time, teachers' own enthusiasm for the garden may simply wane in the face of organisational or practical difficulties – there may be ineffective or no training in how to use the garden in lessons (Blair, 2009) or children's responses in the garden may not be as anticipated. Children are not always universally keen on gardening. Researchers have suggested the importance of accommodating children's designs into wider garden projects (Wake, 2007) and that a garden project may be sustained by children taking more ownership of the garden space. A garden committee, made up of teachers, parents, gardeners and administrators is also proposed (Hazzard, Moreno, Beall, & Zidenberg-Cherr, 2011) in order to maintain children's engagement and let their desires and experiences drive the garden's development.

Volunteers

Volunteers are vital to the long-term running of a school garden. Both New School and City School talked to us about parent-volunteers who came in from the local community to participate in aspects of gardening, and we have heard this in other schools too. However, attracting and maintaining voluntary support is not straightforward. Henryks (2011) has noted how important it is that the needs of volunteers are met; volunteers often become involved in a project for 'one set of reasons and either benefited unexpectedly or gained more than they anticipated' (Henryks, 2011, p. 581). Understanding the underlying motives of volunteers is important to ensure their ongoing engagement with the project. Managing volunteers can become a job in its own right, perhaps an addition to the regular duties of a committed staff member. But staff commitment can wear thin. Lack of staff time alone can make it difficult for schools to have long-term volunteers, or to engage particular members of the local community (Earl, 2018).

These are not the only challenges that schools face in establishing and maintaining a school garden. This is not an exhaustive list. However, the management of challenges, like that of curriculum, depends on how well the school garden is understood as a school change initiative.

The garden as a school change initiative

Our two portraits (Chapters 3 and 4) represent two different approaches to school change – one where the garden is introduced by the head teacher and leadership team as part of a wholistic vision for the school, and the other where the parent body initiates and fundraises for the garden with support from the school leadership team. A third change model, and the one used by the Stephanie Alexander Kitchen Garden programme, is a whole school–sponsored programme involving key teachers and/or teaching assistants and volunteers. A fourth model, that of systemic led and coordinated school change, focusses on aligning activities between various nested layers of activity, beginning at the national level. The Edible School Yard (which can also operate as a single school initiative, the third change model) is integral to California's School Lunch Initiative (SLI). The SLI is reminiscent of earlier US School Garden programmes with its layers of support, regulation and accountability (Barlow & Stone, 2011).

Each of these four approaches has to address key inter-related issues – ownership, capacity and sustainability.

Ownership

For any change initiative to take hold in a school, it must move beyond an enthusiastic few. When education systems want to initiate a universal change, they may just issue a directive, or make the change part of a high stakes audit. However, this is rarely how change happens within a school. An edict from the senior leadership team will generally not suffice.

When an initiative comes from the top, the leader or senior leadership team (SLT) must make sure that it is taken on by staff. School leaders thus:

- offer an inspiring explanation, backed by evidence, of what the change will be and the benefits it will bring
- provide support for carefully managed and evaluated pilots and quick wins
- incorporate the initiative, using early examples, into the school symbolic regime – branding, website, newsletter, prospectus, talks at morning assembly, presentations to governing body – so that the initiative becomes part of the school identity
- take an ongoing personal interest in the progress of the initiative, observing and taking part in activities.

Additionally, the SLT may offer incentives for staff to take up the initiative, such as additional preparation time and/or additional class budget. They also, where practicable, lead by example. This was the approach taken in New School where the initiating head teacher was strongly committed to realising an imaginary of an eco-school through 'green leadership' – for instance, he rode a bicycle to school, personally sourced sustainable materials for new school buildings and undertook a doctorate to further his understanding of practice.

Top-down change initiatives are always stronger and much more likely to succeed if there is already a group of staff – even better with parents and students – who are similarly committed to the same future for the school. The head at New School, for example, involved teachers, students and parents in the design for their new school buildings and yard, and City School gardens were designed in consultation with staff and children. Participatory groups can be 'task and finish' or they can be formalised as an integral part of the development and decision-making processes in the school, monitoring and reporting regularly on change. Such steering groups do run the risk of becoming cliquey, alienating the colleagues they are meant to encourage and support. Top-down change often also struggles against resistant groups of staff, opposition from the community or parents and/or unrealistic expectations about the pace of change. And there are even more challenges if the initiative requires teachers to engage in fundamental pedagogical change which goes against their training and dominant pedagogical practices, as discussed earlier. Nevertheless, leader-initiated school change is widespread and often successful, New School being just one example.

Change initiated from outside the school, as was the case with City School, can be enthusiastically embraced by school leaders, or it can rub up against their priorities for improvement. Most leaders know how to avoid counterproductive friction with well-intentioned and influential parents and community groups – the strategy is usually to accord them a particular role and task, but keep them apart from areas that the leaders want to keep under their direct control. This was certainly the case with City School, where the head's priorities were directed towards shoring up enrolments in a rapidly gentrifying neighbourhood, and becoming a part of the larger vision of the academy chain. Thus, despite having the strong support of the school business manager, the kitchen garden initiative was initially corralled from becoming a whole school change initiative – instead it was seen as a highly beneficial school grounds and fundraising initiative which would support changes in the school meal programme and the outdoor social learning programme. The garden could also be taken up by teachers as they wished and as long it did not detract from maintaining test results in literacy and numeracy.

Change initiated with a strong external organisation and reliant on one or two key staff has variable successes. Such initiatives usually operate on a cascading model of change – there is ongoing expert support from the external organisation for key staff members and they then act as trainers and internal change agents. Senior leaders give support, but often remotely; if they do not incorporate the initiative into the wider school improvement programme, it can founder. But if there is top-level support, good reason and evidence for other teachers to be involved, and incentives for doing so, external organisation–supported programmes can be integrated into everyday school activities. Rather than being a project on the side of the 'real' school programme, the initiatives become integral to school culture and practice. Such is the case with some school garden programmes where they have shifted from being a project to becoming a part of the school's identity and educational programme. The school garden is 'who we are' and 'what we do'.

Integration of a change initiative does not happen overnight, and there are important steps that assist – changing operational and professional capacity is crucial.

Capacity and sustainability

While contemporary schools may not be able to change national testing and inspection regimes, it is still possible to change some of the things that potentially undermine a desired change. Evaluations of kitchen garden programmes often point to difficulties for schools in managing money and time; these can prevent the programme being sustained. It is essential that the composition of the staffing cohort, and the management of time, money and space actively support change rather than impede it. An examination of school management and infrastructure is thus a crucial aspect of capacity-building.

New School adjusted its administrative practice in order to become and be an eco-school. The head appointed a teaching assistant with specific skills in Horticulture and Agriculture. Significant time was allocated each week for her to work with small groups of children tending animals and the allotment, and for doing the preparation required for this educational work to be meaningful and successful. But if responsibility for the success of the programme had been left to this grassroots staff member alone, as is sometimes the case with kitchen garden programmes, eco-school aspirations would not have been realised; that required further action from the top. The New School grounds caretaker also had some oversight of eco-school–related matters, ranging from ensuring that the energy, water and waste systems were working at maximum capacity to supporting the weekly community meals. Eco-school activities were integrated into the school budget and were not a stand-alone item. The annual calendar and weekly timetable included eco-school special seasonal events and regular garden and farm maintenance activities. In contrast, after the first year of operation, City School relied heavily on 'soft money' from fundraising, working bees and volunteer support for the implementation of its kitchen garden programme. The garden was not yet integrated into the regular school time and money resource allocations.

Capacity however is more than administrative structure and resource management. These are necessary but not sufficient. When improvement literatures talk about capacity, they generally refer to professional learning. Professional learning is integral to school change. Building new teacher expertise is not about beginning change but sustaining it. Staff not only need to learn new knowledges and know-how but also often need to *unlearn* old ways of doing things, and/or to redesign existing routines and approaches to accommodate the new.

Our survey of historical school garden programmes shows the kind of centralised support that was provided in the US for teacher learning. External advisers, seasonal calendars, procurement and supply of seed and equipment as well as professional learning – from workshops to graduate certificates in horticulture – were on offer. However, in today's decentralised systems, a comprehensive national offer would likely be considered too intrusive, and insensitive to local needs and decisions. However, if these supports are not provided centrally then they must either be found within the school, or purchased through membership of a network.

Bringing new resources into professional repertoires, testing them out, adapting and evaluating them, is neither a simple nor a quick process. A couple of after-school inputs is hardly sufficient to effect change, as was the case with the introduction of City School's kitchen garden to staff. All of the research on teacher learning and change suggest that a long-term commitment to professional learning is crucial (e.g. Ainscow, 2015; Busher & Baker, 2003; Day, 2007; Harris, James, Gunraj, Clarke, & Harris, 2006; Spillane & Diamond, 2007). Schools often use strategies such as action inquiry and learning sets as a way to support groups of teachers to learn together; New School initially used staff meetings as the means of supporting and sharing classroom-based inquiry. Teachers can also enhance their professional know-how by working away from their schools in intensive immersive experiences and in networks with like-minded practitioners (Lieberman & Wood, 2002, 2003): kitchen garden programmes typically combine this kind of activity together with expert support – and this may morph into the external supporter becoming a critical friend or coach (Swaffield & Macbeath, 2005).

School change is easily undone when staff leave. It is notable that New School was able to retain its focus on ESD when the head teacher changed. This is in part because being an eco-school was the school's identity, and the new head was recruited to lead an ESD-committed school. The successful applicant was chosen to fit the school, and appointed on the basis of their plans to not only maintain but also extend its programmes. However, the garden programme was reduced because of pressures of tests and inspection.

But it is not always the case that replacement heads and teachers are enthusiastic about a predecessor's passion. It is not at all uncommon to go into school yards which bear the signs of abandoned garden projects. New school leaders may decide to divert funds, energy and attention away from the garden into another area they see as more pressing. More commonly, it is the capacity on the ground – literally – that is at issue. When gardens rely almost entirely on one teacher's enthusiasm and leadership, rather than on regular, planned professional learning and sharing of practice, garden programmes almost inevitably falter when they leave. It is important then for leaders committed to garden programmes to engage in realistic succession-planning and risk management in order to ensure that their garden programme does not end up as a few self-seeded potatoes and parsley amid a tangle of weeds.

SLT may also have to attend to some fundamental organisational issues in order to allow teachers to work in new ways. Teachers with enlarged pedagogical repertoires can easily be frustrated by the timetable, organisation of space, and/or lack of equipment or resources. It is thus not at all uncommon for schools pursuing ESD, for instance, to engage with their built environment, and to allow for longer lessons, or a flexible half day where students may engage for longer periods of time with hands-on elective activities.

Back to the question – why garden in schools?

We completed this book during the first COVID-19 lockdown. Like many others, we were more engaged with our own gardens during this period than was usual. We simply had more time. We wanted to be outside and doing something productive. We were drawn to our gardens.

Gardens pull people to them. They entice, they elicit desire, they are perhaps 'vibrant matter' in the sense that Jane Bennett (2010) proposes, non-human actors in an affective ecology which demands particular actions from humans – in this case, appreciative nurture and care. Our gardens strongly reminded us of some of their more intangible and sensual garden joys – the feel of earth on our fingers, seeing seeds sprout before our eyes, watching seedlings lean into the light throughout the day, tasting the first raspberry warmed by the sun. As immigrants to cold Northern climes, we developed more understanding of how to work with plants and bees to achieve germination and fruiting despite the unpredictability of English weather. All children should have these experiences, we think. But schools, of course, have to do more than offer sensual and memorable moments with plants.

The pandemic also heightened our awareness of the fragility of the food supply chain, the importance of farming and the clearer air associated with empty roads. We saw pictures of the Himalayas taken from Kathmandu rooftops, deer roaming deserted city streets, bears playing in empty national park car parks, clear water in Venetian canals. We realised once again how gardening might be part of a curriculum which reoriented children to the natural world – not as apprentices in moral discipline brought about by hard labour (Comenius), nor as wild spirits to be tamed (Rousseau), nor as imperial conquerors of Nature (Herbert Spencer), nor as industrial-scale producers (the fruit orchards of Slovenia). Rather, the garden might be a site for mediating the relationship between children, culture and nature, for meditation on the Anthropocene and its challenges, for considering alternatives beyond techno-driven dystopias and utopias (Diogo, Rodrigues, Simoes, & Scarso, 2019).

As we finished writing, countries were slowly emerging from lockdown and people were beginning to assess how we might live in the future, including how we might educate children well, as well as safely. Commentators like George Monbiot were suggesting that more outdoor, earth-focussed education might be the way forward. In Scotland, authorities had already begun to view outdoor learning in a different light – it offered a template for socially distanced learning. Is now the time for the school garden to become more popular again, we wondered?

We know that there are multiple educational arguments for school gardens. We think that the question of sustainability is one of the most urgent. But regardless of our personal views, we believe strongly that teasing out the various problem–answer rationales for school gardens will help educators think carefully and critically about why they might have a school garden. It is helpful, we think, for schools to ask the same question we used – 'What's the problem the garden will answer?' – to begin a garden-planning process. The three areas outlined in this chapter on curriculum connections, practical challenges and change processes might then inform the development of detailed action plans by schools wishing to establish a garden, or guide critical evaluations of existing school garden programmes.

But we think it is also helpful to ask counter questions. As we look back on the book, we are struck by the weight of the expectations that have been made of the school garden. We ask ourselves, do we ask too much of school gardens? Is it seeing the garden as a site for multiple and various learnings actually a weakness rather than a strength? Would a singular focus be better? What can a garden do and what cannot it do? What purpose or purposes is the school garden best suited for? Can a garden be less tied down, more open, more fluid, more amenable to becoming teacher?

We hope that our writing has helped you to think about your views and hopes for school gardens. We think it is useful to conceive of and define school gardens in the widest possible way. Moving from the notion of a school garden as a space for healthy eating or knowing where your food comes from, to a space where anything can happen and anything is possible. Such an understanding of school gardens as more than edible spaces aimed at crafting healthy citizens might, we hope, allow children and adults to experience them differently. But this is just our view. We look forward to hearing what you think.

Postscript

At the time of writing this book, parents were not running their bi-annual food festival at City School, although the school website indicated that the festival would return in 2021. The head teacher had written to parents mentioning the school's well-established 'wonderful garden', and a brief description on the website (on fundraising) spoke of a new eco-cabin, recycling area and a firepit in the bamboo area, in addition to the garden. The most recent Ofsted report indicated that children loved the outdoor-learning activities because these gave them 'memorable experiences' like fire-building. The garden was noted as being used to grow vegetables and herbs for the school kitchen. It appears that the garden occupies much the same place in the school as it did when we visited some years previously. The garden was an asset, connected to food knowledge and healthy eating, a site for some outdoor learning, but linked variably to the wider curriculum. Learning through the garden had, we guessed, not become a key part of the school's educational identity: this was still geared to resilience combined with learning attainment and excellence.

New School improved achievements in mathematics and literacy from Year Three onwards but was still struggling to meet national averages in the early years. Eventually, it became a stand-alone academy with new governance, school uniform, logo, name and mission. While some farm school and garden activities have continued, it is now no longer an eco-school. Most recently it has become leader of a local multi-academy trust committed to improving community relations and attainment – the head teacher from 2012 now leads the trust, and the curriculum is standardised throughout. We get the impression that while the farm and garden are not actively promoted, they are still, however, an important part of the school's programme – just as we might have predicted when we last visited.

Questions to ask when setting up a school garden

Purposes

What is the problem for which the garden is the answer?
How do we know this is a problem? What evidence do we have?
What other answers might there be to this problem than a garden?
Is the garden then the best solution? Why?

Starting the school garden

Support

Who needs to support the idea of a school garden? Parents? Children? Staff? Community? What evidence will they need?
How will we communicate this to them?
How will we keep them informed?
Can we involve them in the garden? How? Do we need someone to coordinate volunteer involvement?

Space

How big will the garden be?
Do we have the land required?
What will be lost if we have a garden?
What condition is the land in?
What needs to be done in order for it to become a garden?

People

Who will do the initial work on making the land ready for a garden? When?
Who is going to establish the garden?
What do they need to know and be able to do in order to establish the garden?
Where can they learn this if they don't already know?
Who will look after the garden on a regular basis? What will we do in the school holidays?
Does the staffing complement of the school need to change to accommodate the garden?

Money and time

How long will it take to do the initial preparation work? Where is this time to come from? How will it be paid for? How much will it cost?

What equipment and facilities do we need? How much will they cost?
How much do we budget for plants and compost to start with? And from then on?
How will the garden be regularly timetabled?

Curriculum

Where and how is the garden to be connected to the curriculum?
Who can learn what in the garden, how often and when?
What support is needed to develop a garden curriculum?
What initial learning do teachers need? What ongoing support?
Where will this come from? Is there an external organisation that can help? How much will this cost?
Are there local, Indigenous and community knowledges about gardens we can draw on? How and where?
How will we phase in garden learning? What comes first?

Changing what we do

What is our theory of change? What do we all need to learn together about change? What kind of learning communities do we need? Who will be involved? How will these be supported and by whom?
What school practices might hinder the garden – discipline policy, pedagogical practices, beliefs about children's development and capabilities? Assessment, esting and exam practices?
How can we learn from other schools' experiences with gardens?
How will we evaluate the garden? And how often?
What will count as 'success'?
How will we make decisions about what to do in and for the garden? Who needs to be involved? How does this connect to our school governance structure?

References

Ainscow, M. (2015). *Towards self-improving school systems: Lessons from a city challenge*. London: Routledge.

Ball, S. (2018). *The education debate* (3rd ed.). Bristol: The Policy Press.

Barlow, Z., & Stone, M. K. (2011). Living systems and leadership: Cultivating conditions for institutional change. *Journal of Sustainability Education, 2*(March), 1–29.

Bennett, J. (2010). *Vibrant matter. A political ecology of things*. Princeton, NJ: Princeton Unviersity Press.

Bittner, W. (1919). *Public discussion and information service of university extension. Bulletin 61.* Washington, DC: Department of the Interior, Bureau of Education.

Blair, D. (2009). The child in the garden: An evaluative review of the benefits of school gardening. *The Journal of Environmental Education, 40*(2), 15–38.

Boyle, B., & Bragg, J. (2008). Making primary connections: the cross-curriculum story. *The Curriculum Journal, 19*(1), 5–21.

Busher, H., & Baker, B. (2003). The crux of leadership. Shaping school culture by contesting the policy contexts and practices of teaching and learning. *Educational Management and Administration, 31*(1), 51–65.

Datnow, A., Hubbard, L., & Mehan, H. (2002). *Extending educational reform: From one school to many.* London: Routledge Falmer.

Day, C. (2007). Sustaining the turnaround: What capacity building means in practice. *ISEA, 35*(3), 39–48.

Day, C., Gu, Q., & Sammons, P. (2016). The impact of leadership on student outcomes: How successful school leaders use transformational and instructional strategies to make a diffference. *Educational Administration Quarterly, 52*(3), 221–258.

Diogo, M. P., Rodrigues, A. D., Simoes, A., & Scarso, S. (Eds.). (2019). *Gardens and human agency in the Anthropocene.* New York: Routledge.

Earl, L. (2018). *Schools and food education in the 21st Century.* Abingdon: Routledge.

Earley, P., & Greany, T. (Eds.). (2017). *School leadershp and education system reform.* London: Bloomsbury.

Elmore, R. (2004). *School reform from the inside-out.* Cambridge, MA: Harvard University Press.

Fullan, M. (2011). *Change leader: Learning to do what matters most.* San Francisco, CA: Jossey Bass.

Galton, M., Hargreaves, L., & Pell, T. (2009). Group work and whole-class teaching with 11–14 year-olds compared. *Cambridge Journal of Education, 39*(1), 119–140.

Hall, C., & Thomson, P. (2017). *Inspiring school change. Transforming education through the creative arts.* London: Routledge.

Hargreaves, A. (Ed.) (1996). *Rethinking educational change with heart and mind.* Alexandria, VA: Association for Supervision and Curriculum Development.

Hargreaves, A., & Goodson, I. (2006). Educational change over time? The sustainability and nonsustainability of three decades of secondary school change and continuity. *Educational Administration Quarterly, 42*(1), 3–41.

Harris, A., James, S., Gunraj, J., Clarke, P., & Harris, B. (2006). *Improving schools in exceptionally challenging circumstances. Tales from the frontline.* London: Continuum.

Hayes, D., Mills, M., Christie, P., & Lingard, B. (2005). *Teachers and schooling: Making a difference. Productive pedagogies, assessment and performance.* Sydney: Allen & Unwin.

Hazzard, E., Moreno, E., Beall, D., & Zidenberg-Cherr, S. (2011). Best practices models for implementing, sustaining, and using instructional school gardens in California. *Journal of Nutrition Education and Behaviour, 43*(5), 409–413.
Henryks, J. (2011). Changing the menu: Rediscovering ingredients for a successful volunteer experience in school kitchen gardens. *Local Environment, 16*(6), 569–583.
Lieberman, A., & Wood, D. (2002). *Inside the national writing project: Connecting learning and classroom teaching.* New York: Teachers College Press.
Lieberman, A., & Wood, D. (2003). Sustaining the professional development of teachers: Learning networks. In B. Davies & J. West-Burnham (Eds.), *Handbook of educational leadership and management* (pp. 478–490). Edinburgh Gate: Pearson Longman.
Lingard, B., Christie, P., Hayes, D., & Mills, M. (2003). *Leading learning: Making hope practical in schools.* Buckingham: Open University Press.
Macbeath, J., & Dempster, N. (2009). *Connecting leadership and learning. Principles for practice.* London: Routledge.
Mayall, B., & Morrow, V. (2011). *You can help your country: English children's work during the Second World War.* London: Institute of Education.
Mills, M., & McGregor, G. (2016). Learning not borrowing from the Queensland education system: lessons on curricular, pedagogical and assessment reform. *The Curriculum Journal, 27*(1), 113–133.
Priestley, M., & Sinnema, C. (2014). Downgraded curriculum? An analysis of knowledge in new curricula in Scotland and New Zealand. *The Curriculum Journal, 25*(1), 50–75.
Salter, P., & Maxwell, J. (2016). The inherent vulnerabiity of the Australian Curriculum's cross-curriculum priorities. *Critical Studies in Education, 57*(3), 296–312.
Sipe, S. (1909). *School gardening and nature study in English rural schools and in London.* Washington, DC: US Department of Agriculture.
Spillane, J., & Diamond, J. B. (Eds.). (2007). *Distributed leadership in practice.* New York: Teachers College Press.
Swaffield, S., & Macbeath, J. (2005). School self-evaluation and the role of a critical friend. *Cambridge Journal of Education, 35*(2), 239–252.
Taylor, C. (2016). Implementing curriculum reform in Wales: The case of the Foundation phase. *Oxford Review of Education, 42*(3), 299–315.
Thomson, P. (2011). *Whole school change. A reading of the literatures* (2nd ed.). London: Creative Partnerships, Arts Council England.
Thomson, P., Jones, K., & Hall, C. (2009). *Creative whole school change. Final report.* London: Creativity, Culture and Education; Arts Council England..
Thomson, P., Hall, C., & Jones, K. (2012). Creativity and cross-curriculum strategies in England: tales of doing, forgetting and not knowing. *International Journal of Educational Research, 55*, 6–15.

Thomson, P., & Riddle, S. (2019). Who speaks for teachers? Social media and teacher voice. In A. Baroutsis, S. Riddle, & P. Thomson (Eds.), *Education research and the media: Challenges and possibilities.* (pp. 119–134). London: Routledge.

Tyack, D., & Tobin, W. (1994). The grammar of schooling: Why has it been so hard to change? *American Educational Research Journal, 31*(3), 453–480.

Tyack, D., & Cuban, L. (1995). *Tinkering toward utopia. A century of public school reform.* San Francisco, CA: Jossey Bass.

Wake, S. J. (2007). Children's gardens: Answering 'the call of the child'? *Built Environment, 33*(4), 441–453.

Wrigley, T., Thomson, P., & Lingard, B. (Eds.). (2011). *Changing schools: Alternative approaches to make a world of difference.* London: Routledge.

Index

Ackley, C. 46
Adams, J. 48
Adatia, R. 94
adminstration: school 123–124
Agamben, G. 8
agency: of children 71–74
agriculture 3, 28
Ainscow, M. 112, 124
Almers, E. 49
Altenberger, E. 47
Amanti, C. 48
Armitage, K. C. 27
Art 70–71
assemblage 7–8, 11; discursive 102–104

Bacchi, C. 4–6
Bailey, L. H. 27
Baker, B. 124
Ball, R. 94
Ball, S. 7, 39, 48, 101, 116
Baquedano-Lopez, P. 48
Barale, K. 94
Barber, M. 39
Barry, D. 45
Bartosh, O. 48
Beales, B. 4, 78
Beall, D. 120
Beery, M. 94
Begley, P. 46
Bell, A.C. 47
benefits of gardens 21, 40–44
Bennett, J. 125
Bentsen, P. 100
Bernstein, B. 48
Bethel, A. 39
Birdshall, S. 48
Bittner, W. 113
Blair, D. 120
Block, C. 94
Block, K. 42, 43
Bloom, J.D. 45
Bolling, M. 100
Bonham, J. 4
Bourdieu, P. 95
Bowers, C.A. 49
Boyle, B. 116
boys *v* girls 24
Bragg, J. 116
Braun, A. 48
Breeze, T. 48
Brehony, K. 15
Brewer, G.W.S. 25, 26
Briggs, J.M. 94
Brown, J.D. 96
Burke, C. 35, 96
Burt, K.G. 20
Busher, H. 124

Cairney, P. 39
Cairns, K. E. 95, 106
Camic, P. 47
Canadian Heritage Matters 28
Canaris, I.S. 94
capacity building 123–125
Carden, A.M. 94
Carpenter, L. 43
Carson, R. 27
central support 124
challenges 43–44; funding 63–67; for teachers 116
Chan, T.C. 46

child development 17
child-centred learning *v*. book learning 19
children: agency 98; influence on family 94; as outputs of garden programmes 95
Christie, P. 112, 117
Christodolou, A. 47
Chung, K. 48
citizenship 106
city: as corrupt 23; *v* suburbs 24; as ugly 23
City School: context 60; school change 62–64
civilising, garden as 106
Clapp, H. 23
Clarke, P. 124
class 36, 95, 105–106
Cole, L.B. 47
Collins, C. 94
Comber, B. 4
Comenius, J. A. 13–14, 126
Community of Gardens 20
Condon-Paoloni, D. 42
conservation 27
control *v*. freedom 71–74
cooperation 106
core skills 89
Crawford, P. 41
critical friend 125
cross-curriculum 114–117
Cuban, L. 117
cultivation 28
curriculum 25–29; connections with gardens 85–87, 103–104, 112–114; frameworks 115; integration of gardens into 48, 69–70; materials 115; system support for 115; and teachers 47–49
Curtis, R. 4, 8, 78
Cutter-Mackenzie, A. 98

D'Abundo, M.L. 94
Dahlberg, G. 100
Darwin, C. 16
Datnow, A. 112
Davies-Barnes, S. 48
Davies, H. T. O. 39
Davis, J.N. 94
Dawber, J. 42
Day, C. 4, 112, 124
deficit discourse 105–106
Deleuze, G. 7
Dempster, N. 112
Desmond, D. 20
developmental discourse 100
Dewey, J. 18–19
Diamond, J.B. 124
Diaz, J.M. 45, 47
Dimbleby, H. 94
Diogo, M.P. 126
dirt 75
disciplines 21
discourse 4–5; deficit 105–106; developmental 100; silences 104–106
Dowler, E. 96, 97
Dudek, M. 35
Duhn, I. S. 99
Dyg, P.M. 47
Dyment, J. 47

Earl, L. 4, 36, 37, 69, 81, 93, 95, 105, 120
Earley, P. 112
Eckermann, S. 42
eco-building 79–80
eco-school 45; vision for 78–79
Edible School Yard 40–41, 68, 94
Edmundsen, J. 35
Edmundson, D. 49
Education for Sustainable Development 102, 113
educational historians: approach to gardens 34–36
educational psychology 16
Elkadi, H. 47
Elmore, R. 112
Emerson, P. 26
enthusiasm for garden 119
environment 36–38; crisis of 102
evaluation what works approach 39–40
experiential learning 14–19; direct instruction and 16, 18

Fakharzadeh, S. 41
family, children influence eating habits of 94
farm education 80–82
Ferguson, B.G. 48

Fildes, D. 42
Flett, M. 94
Flowers, R. 95
food: discourses 93; education 35, 36–38, 68–69, 82–83, 87, 95; hunger 6–61, 96; provenance 68, 101; waste 75
Food Festival 63–67
Food for Life 83
foodie discourse 40
foodieness 92–97, 102–104
Forrest, M. 35
Foucault, M. 4–5, 7, 36
freedom 99–101
French, A. 28
Froebel, F. 15–16, 34
Fullan, M. 112
funding 63–67, 118
Furhman, N.E. 47

Gabler, E. R. 28, 29
Galton, M. 116
Gard, M. 105
garden: challenges of 118–120; as civilising 17, 18, 21, 106; and class 36; as colonising 28; as cross-curriculum project 114–117; and curriculum 70, 85–87; and curriculum connections, multiple 25–29, 112–114; and curriculum integration 103–104; design of 14–15, 26; enthusiasm for 119; former 59; funding for 118; and gender 35; good for troubling and troublesome children 24; histories of 34–36; outcomes, measurable 98; maintenance 119; moral learning 22, 25, 35; as multiple 2–3; multiple places 82; as orderly 74–75; outdoor learning 64; positive benefits 21; as practical skills 24; restricted access 99–100; as safe space 74, 75; as school change 121; seasons 119; as separate subject 26, 112–113; space 118; as vibrant matter 125–126; vocational learning, site for 21–22; volunteers 119
garden programmmes: address multiple problems 23; Australia 34; benefits of programmes 40–44; challenges of 43–44; China 35; cultural assumptions of 105; and conservation 27; Ireland 34; India 35; during World Wars 20; mixed mandates 20–21; multiple purposes 20–22; and nationalism 34–35; outcomes hard to assess 98; Prussia 34; social context, importance of 36; Slovenia 34; solve problems of city 23
Gargano, E. 17
Garside, R. 39
Garthwaite, K. T. 96
Gaylie, V. 35, 47, 98
Geissman, R. 94
gender 35
Gentry, S. 39
Gibbs, L. 42, 43, 94
girls *v* boys 24
Gold, L. 42
Gonzales, N. 48
Goodson, I. 112
Gough, A. 47
Gough, D. 39
grammar of schooling 117
Greany, T. 112
Gree, A. 45
green leadership 46, 122
Green, M. 99
Greene, L. 20, 21
Grieshop, J. 20
Grootemaat, P. 42
Gu, Q. 112
Gudmund-Hoyer, M. 7
Guraj, J. 124
Guthman, J. 36, 62, 76, 95, 105

Hacking, I. 101
Hall, C. 4, 45, 48, 112, 116
Hamilton, L. 98
Harbor, J. 26
Hargreaves, A. 112
Hargreaves, L. 116
Harris, A. 124
Harris, B. 124
Hayden-Smith, R. 20, 23, 35
Hayes-Conroy, J. 37, 95
Hayes, D. 112, 117
Hayman, G. 43
Hazzard, E. 120

health: crisis 92–93; education 36–38; promotion 94
healthy eating 40, 93–95
Hemenway, H. D. 13, 24, 25
Henderson, C. R. 94
Henryks, J. 46, 120
Herington, S. 15, 34
Hinds, J. 47
Hinrichs, C.C. 45
Hipkiss, A.M. 49
Hoffman, A. 35
Holmes, K. 34
horticulture 3, 25–26, 28
Howard, S. 100
Hubbard, L. 112
human-centrism 35, 97–98
Hunter, D. 47
Hunter, I. 74
Hyde, W. 48

independence 73
indigenous knowledges 34, 115
Ingram, V. 35
inspection 116
Izadpanahi, P. 47

Jaenke, R.L. 94
James, S. 124
Jarvis, C.D. 24
Jekyll, G. 36
Jewell, J. R. 17
John, V. 48
Johnson, B. 42, 43, 94, 100
Jones, K. 4, 112, 116
Jones, R. G. 25
Jones, S. 48
Jorgenson, S. 101

Kadji-Beltran, C. 45
Kass, D. 27
Kelly, P. 69
Kelly-Richards, S. 98
Kemp, N. 48
Kensler, L.A.W. 46
Kiddle, R. 48, 49
Kihn, P. 39
Kincy, N. 47
Kirk, D. 48
Kjellstrom, S. 49
Knauft, D. 47
Knausenberger, A. 45
Kneen, J. 48
Knight, S. 99, 100
knowledge: community funds of 24; disciplinary 21; experiential *v* disciplinary 17–19; funds of 48; Indigenous 34, 24, 115; local funds of 41; Travellers 86, 88
Koch 94
Kohlstedt, S. 20, 27
Kopsell, D. 45
Korfiatis, K. 47
Kulas, J. 42

Laeyva-Cutler, B. 48
Laird, S. 41
Lambie-Mumford, H. 96, 97
Lashey, L. 46
Lather, P. 39
Latter, L. 17, 21, 23, 24, 26
leadership 46, 78–79; changes in 87–91
learning: book 18; book *v*. child centred 19; child centered 19; as developmental 21, 100; inner need for 15; moral 15; outdoor 64, 126; practical skills 24; social and emotional 17; through experience 14–15, 17, 19; through senses 14–15; vocational 21–22
a learning garden 67–68
Lee, J. C-K. 47
Lee, K. 48
Legg, S. 7
Leo, U. 46
Let's Move 94
Lieberman, A. 124
Lineberger, S.E. 94
Lingard, B. 112, 117
Livermore, A. L. 21
Loftus, L. 45
Logan, A. 2, 25, 28
Long, C. 42
Loopstra, R. 96
Louv, R. 74, 101
Lovell, R. 39
Loxley, J. 39
Lubans, D.R. 94
Lupinacci, J. 49
Lupton, D. 105

Macbeath, J. 112, 125
Macfarlane, S. 42

"magic carrot" model 95
Maguire, M. 48
Mahatma Ghandi 35
maintenance of garden 119
MalbergDyg, P. 98
Mangual Figueroa, A. 48
Mann, A. 49
Mannion, G. 49
Mao Zedong 35
maps 13
Markwell, K. 94
Marston, S.A. 98
Martin, S. K. 34
Marturano, A. 20
Martusewicz, R. 49
Mawer, C. 48, 49
Mayall, B. 35, 112
Mayer-Smith, J. 48
McAuliffe Bickerton, C. 48, 49
McFarlane, S. 94
McGregor, G. 117
McQuade, V. 4
Mehan, H. 112
Meyers, B. 94
Miller, L. 24, 25, 26
Mills, M. 112, 117
Minton, T.G. 27
Mirmohamadi, K. 34
models of school change 121
Moffit, A. 39
Moll, L. 48
Monbiot, G. 126
Montessori, M. 17–18, 21
Moore, S. A. 98
moral learning 15, 22, 25, 35
Morales, H. 48
Moreno, E. 120
Morgan, P. 94
Morris, D. 42
Morris, J. 94
Morris, M. 44, 72
Morrow, V. 35, 112
Moss, P. 100
Murphy, J. M. 41
Mycock, K. 49, 99
Mygind, E. 100

Narayan, E. 48
nationalism 34–35
Nature: appreciation of 17; and cultivation 28; deficit 61–62, 97; History 17; and human-centrism 35; human interaction 97–98; human relations, reorientation of 12; taming 16; as teacher 15
Nature Study 26–29, 70, 113; movement 27
Navarro, M. 47
New School context 77–80
Nielsen, W. 42
Nigh, R. 48
Nnakwe, N. 45
nostalgia 101
Nutley, S. M. 39
nutrition 41

O'Connor, D. 96, 97
O'Kane, F. 34
O'Koumunne, O. C. 42
obesity 6–7, 93–94, 104–105
Ofsted 87–88, 89, 127
Ohly, H. 39
Oliver, K. 39
Oliver, S. 39
Oswald, M. 100
Outdoor Education 100
outdoor learning 99–101, 126
ownership 121–123

Parsons, H. 21–22
Passy, R. 44, 48, 99
Patrick, R. 96
Pell, T. 116
Peltonen, M. 7
Pence, A. 100
Pennacchia, J. 4
Per, A. 49
performativity 39
Personal Social Health Education (PSHE) 114
Pestalozzi, J-H. 14–15
Peterat, L. 48
physical activity 23–24
Pike, J. 69
plots individual *v*. communal garden 26
policy framing 48
Potter, L. 101
poverty 35, 60
poverty 60
Priestley, M. 114
problematisations 101

professional development 67
professional learning 124–125
progressivism 19, 41
provenance of food 68
Pudup, M.B. 76

Quinsey, K. 42

Raffinsoe, S. 7
Rainville, K. 45
Randomised Control Trial 39
Randrup, T. 100
Rauzon, S. 41
Reed, F. 44
resilience 73, 100–101
resources, management of 13–124
Ribaric, M. 34
Riddle, S. 116
risk-taking 72, 100
Roberts, A. 47
Robin, L. 34, 101
Rochford, K. 4
Rodrigyes, A.D. 126
Rogers, E.M. 47
Rossi, T. 48
Rousseau, J-J. 14, 126
Royal Horticultural Society 43–44
Ryan, A. 48, 49

Sammons, P. 112
Sanders, D. 49
Sandolod, C. 45
Saunders, K.L. 94
Saunders, R. 46
Savage, M. 7
Scarso, S. 126
Schafft, K. 45
school change 16–17, 45–47, 121–123; administrative practice 123–124; central support for 124; critical friend 125; management of resources 123–124; models 121; outside initiation 122–123; stakeholders, mixed views 62–64; sustaining 45–46, 123–125; top down 122; as vernacular 45; with external partner 123
school dinners 96–97
school discipline 71, 100–101
School Food Plan, The 93–94
school grounds 47
school subjects 47–49
schooling grammar of 117
Schwab, E. 17–18, 26
science 17, 25, 70
seasons, influence on planting 119
Segantin, O. 94
senses 14, 15
Siemonek, L. 48, 49
Silvasti,T. 96, 97
Simoes, A. 126
Sinnema, C. 114
Sipe, S. 23, 118
Skaer, C.F. 94
Smith, D.I. 14
Smith, P. C. 39
social and emotional learning 17
social context, importance of 36
Somerset, S 94
space garden 118
Spalding, A.D. 45
Spaniol, M.R. 94
Spencer, H. 16, 126
Spillane, J. 124
staff changes 125
Staiger, P. 42, 43, 94
Stapleton, S.R. 46
Steffen, R. 45
Stephanie Alexander Kitchen Garden 41–42, 48, 113, 115; benefits of 41–42; problems 41–42
Stevenson, R.B. 45
Stone, M.K. 47
Studer, N. 41
subjects, garden connections with 69–70
Subramaniam, A. 20
Surman, E. 98
Swaffield, S. 125
Swan, E. 95
system support 115
systematic review 39

taste 95
Taylor, C. 114
teacher learning 67, 124–125; garden knowledge 17
teachers and change 47–49; challenged by change 116; and curriculum 47–49; *v.* garden assistants 26
testing 116
Thaning, M. S. 7

Thayer, E. 48
Thomas, J. 39
Thomas, U. 47
Thomson, P. 4, 45, 48, 78, 112, 116
time 117
Tiplady, L. 47
Tobin, W. 117
Todd, L. 94
Townsend, M. 42
Traveller food traditions 86, 88
Trelstad, B. 23, 35
troubling and troublesome children 24–25
Tsang, E P.C. 47
Tucker, R. 47
Tyack, D. 117

Ulme, C.L. 46
United States Department of Agriculture 20
United States Garden Army 21, 25, 26
unlearning 124
US School Garden history 20–29; influence of European pedagogues on 27

Veronese, D.P. 46
vibrant matter 125–126
Vincent, J. 93
vocational learning 21–22
volunteers 46, 119

Waggoner, E. 24, 26
Waite, S. 100
Wake, S.J. 101, 120
Waliczek, T.M. 94
Wang, M. 41
Warner, L. A. 45
Waters, A. 45, 68
Waters, E. 42
Webb, S. 45
Weed, C. M. 26
Wells, M. 4
Wells, N.M. 94
Westall, C. 101
what's the problem approach 5, 7
White House Kitchen Garden 94
Whitehead, K 34
Whiteness 95
Wickenberg, P. 46
Wigglesworth, R. 39
Williams, D.R. 96
Wilson, J. 98
Windsor, S. 49
Wistoft, K. 47, 48, 98
Wood, D. 124
Woolner, P. 47
World Health Organisation: Department for Health 7
World Wars 20, 28, 35, 101, 113
Wright, J. 105
Wrigley, T. 112

Yeatman, H. 42, 47
Young, D. 43

Zachariah, M. 35, 45
Zajicek, J. M. 94
Zidenberg-Cherr, S. 94, 120

Lightning Source UK Ltd.
Milton Keynes UK
UKHW050734130322
399787UK00021B/89